Trust in the system

Manchester University Press

INSCRIPTIONS

Series editors
Des Fitzgerald and Amy Hinterberger

Since the very earliest studies of scientific communities, we have known
that texts and worlds are bound together. One of the most important
ways to stabilise, organise, and grow a laboratory, a group of scholars,
even an entire intellectual community, is to write things down. As
for science, so for the social studies of science: Inscriptions is a space
for writing, recording, and inscribing the most exciting current work in
sociological and anthropological – and any related –
studies of science.
The series foregrounds theoretically innovative and empirically rich
interdisciplinary work that is emerging in the UK and internationally. It is
self-consciously hospitable in terms of its approach to discipline (all areas
of social sciences are considered), topic (we are interested in
all scientific objects, including biomedical objects), and scale (books
will include both fine-grained case studies and broad accounts of
scientific cultures).
For readers, the series signals a new generation of scholarship captured
in monograph form – tracking and analysing how science moves through
our societies, cultures, and lives. Employing innovative methodologies for
investigating changing worlds, Inscriptions is home to compelling new
accounts of how science, technology, biomedicine, and the environment
translate and transform our social lives.

Trust in the system

Research Ethics Committees and the regulation of biomedical research

Adam Hedgecoe

MANCHESTER UNIVERSITY PRESS

Published by Manchester University Press
Oxford Road, Manchester M13 9PL

www.manchesteruniversitypress.co.uk

British Library Cataloguing-in-Publication Data
A catalogue record for this book is available from the British Library

ISBN 978 1 5261 5291 6 hardback
ISBN 978 1 5261 6705 7 paperback

First published 2020
Paperback published 2022

Typeset by Newgen Publishing UK

To Sheridan, with love.

Contents

Acknowledgements

I am grateful to the European Commission (grant number MEXT-CT-2003-509551-ED-REG-HAR) for funding the research this book draws on and to John Abraham for helping me think about the original grant application.

I am extremely grateful to the Research Ethics Committee members who agreed to be observed and everyone who allowed me to interview them. The nature of presenting ethnographic data in an ethical, anonymised fashion necessarily means these people, without whom this book would not exist, never get the public recognition they deserve.

Richard Ashcroft and Naomi Pfeffer were both vital in guiding my initial thoughts about this topic, Professor Pfeffer most obviously through giving me access to her personal archive of policy documents.

Over the years at Cardiff University, a number of colleagues have provided me with thoughtful feedback on my ideas. I would like to thank Harry Collins, Rob Evans, Nicky Priaux, Martin Wienel, Gareth Thomas, and Joanna Latimer for their time and help. I would also like to thank Laura Stark and Chris Goldsworthy for their readings of the whole manuscript and suggested improvements.

Some of the data and ideas in this book have previously been published in:

- 'Scandals, ethics and regulatory change in biomedical research', *Science, Technology, & Human Values*, 42:4 (2017), 577–599.
- 'Trust and regulatory organizations: The role of local knowledge and facework in research ethics review', *Social Studies of Science*, 42:5 (2012), 662–683.

- 'A deviation from standard design? Clinical trials, research ethics committees and the regulatory co-construction of organizational deviance', *Social Studies of Science*, 44:1 (2014) 59–81.

I am making final revisions to this book in April 2020, at the height of the coronavirus lockdown. I would like to thank Sheridan, Davis, and Jed for supporting me in this whole project but especially in the final stages, at such a strange time.

Abbreviations

ABPI	Association of the British Pharmaceutical Industry
AMS	Academy for Medical Sciences
BMA	British Medical Association
BMAA	BMA Archive
CCHMS	Central Committee for Hospital Medical Services
CEC	Central Ethical Committee
CHCs	Community Health Councils
CNEP	Continuous Negative Extrathoracic Pressure
COREC	Central Office for Research Ethics Committees
CRO	contract research organisation
CRS	cytokine release syndrome
CTIMPs	Clinical Trials of Investigational Medicinal Products
DER	Designated Engineering Representative
DH	Department of Health
DHSS	Department of Health and Social Security
DSMB	Data Safety Monitoring Board
FAA	Federal Aviation Administration
FTIM	first time in man
GAfREC	*Governance Arrangements for NHS Research Ethics Committees*
GMC	General Medical Council
HRA	Health Research Authority
IB	Investigator's Brochure
ICH	International Conference on Harmonisation
IRB	Institutional Review Board
LREC	Local Research Ethics Committee
MAP	morning-after contraceptive pill

MASC	Medical Academic Staff Committee
MHRA	Medicines and Healthcare products Regulatory Agency
MMR	measles, mumps, and rubella
MRC	Medical Research Council
MRECs	Multi-centre Research Ethics Committees
MRI	magnetic resonance imaging
NA	National Archives
NBS	National Blood Service
NHS	National Health Service
NPSA	National Patient Safety Agency
NREAP	National Research and Ethics Advisors' Panel
NRES	National Research Ethics Service
PHS	Public Health Service
PI	principal investigator
PIS	participant information sheet
PK	pharmacokinetics
PR	public relations
R&D	Research and Development
RCP	Royal College of Physicians
RCPA	Royal College of Physicians Archive
RECs	Research Ethics Committees
ShED	Shared Ethical Debate
SOPs	standard operating procedures
STS	Science and Technology Studies
UCH	University College Hospital
VC	venture capitalist

Introduction – On the margins of a trusting system

'Distrust is something which takes place on the margins of trusting systems.'

Steven Shapin[1]

'Often, then, a scientific community has no alternative to trust, including trust in the character of its members.'

John Hardwig[2]

Every month, all over the UK, groups of men and women from a range of backgrounds – doctors, biomedical researchers, nurses, and non-medically trained lay colleagues – sit in meetings, behind piles of paper, and make decisions that affect millions of people. These decisions – about what research should be carried out within the UK National Health Service (NHS) – help determine what drugs come to market, what techniques psychiatrists use to treat mental illness, and the success of any number of other medical interventions. These groups, known collectively as Research Ethics Committees (RECs), are charged with protecting the interests of patients who may be enrolled in these projects and ensuring that the research is carried out to acceptable ethical standards. They review research prior to it taking place, and approval from one of these committees is a requirement before research starts. Carrying out some kinds of research – pharmaceutical trials – without such approval is illegal and, for some professions, research without approval can be a career-ending mistake. For example, while Dr Andrew Wakefield gained notoriety for his 1998 study putatively linking the measles, mumps, and rubella (MMR) vaccine to cases of autism, he was struck off the UK medical register in 2010, in large part for failing to keep to the terms of his REC approval.[3]

Typically, these longstanding, powerful groups maintain a comparatively low public profile. On occasion, for example following the disastrous 2006 first-in-human trial of a new drug at Northwick Park Hospital in London, RECs do receive a degree of press scrutiny, but more often than not, they are simply mentioned as part of the general background to medical research. In contrast, within the medical literature – especially general-interest medical publications such as the *British Medical Journal* – RECs have a high (if unwelcome) profile. The shelf above my desk includes three lever-arch files, bulging with articles and letters to the editor discussing – i.e. complaining about the iniquities of – REC review of medical research in the UK. Next to them are two similar files with papers about such review in the US. Given that NHS RECs, far from being imposed on the medical profession from outside, began – and indeed remain to a large extent – a form of professional self-regulation, the first aim of this introductory chapter is to explore the paradox of these complaints.

In broad terms, the approach offered in this book aims to do for research ethics review what Steven Shapin has done for modern science, and unpack an ostensibly impersonal, de-socialised aspect of late modern culture, revealing its highly personal, subjective, and social nature. This broad context is an important one, in that: 'What is modernity, and even more its "late" version, but the subjugation of subjectivity to objectivity, the personal to the methodologically mechanical, the individual to the institutional, the contingent and the spontaneous to the rule of rule?'[4] As part of this programme, Shapin places considerable emphasis on the role of trust relations in the creation and maintenance of scientific knowledge, pointing out that:

> Science is a trusting institution ... It is only by trusting others that scientists hold the vast bulk of their knowledge, that their knowledge has scope, that they can know things they themselves have not experienced, and, indeed, that they can be effectively skeptical when they wish to be.[5]

The basis for such trust relations varies over time. In *A Social History of Truth*, a detailed history of the origins and development of the Royal Society of London in post-Restoration England, Shapin argues for the importance of 'some kind of moral bond between the individual and other members of the community' to make 'knowledge ... effectively accessible to an individual'.[6] In this case the 'moral bond' (trust) was intimately connected with one's social status: in essence one was trustworthy (and hence a good scientist) if one was a 'gentleman'.

More recently, Shapin has explored twentieth-century technoscience, including such late modern configurations as the 'industrial research laboratory and the entrepreneurial network', arguing that personal virtue, familiarity, and charisma are not remains from some pre-modern proto-science – 'We are not here talking about premodern survivals or vestiges' – but rather (because of the proliferating uncertainties of late modernity) at the core of the scientific life: 'The closer you get to the heart of technoscience, and the closer you get to the scenes in which technoscientific futures are made, the greater is the acknowledged role of the personal, the familiar and even the charismatic.'[7]

While following a similar trajectory, this book's aims are, almost inevitably, more modest: to explore the lived reality of REC decision making – a process typically presented as impersonal, methodologically mechanical, and in service to the 'rule of rule' – painting a picture that emphasises the central role of trust in such decisions, their local, embedded nature, and the importance of personal knowledge of researchers. The central insight of this book is that, to paraphrase Shapin, the closer you get to the heart of research ethics review, the greater is the acknowledged role of the personal, the familiar, and even the charismatic.

In presenting RECs as a form of regulator, I am acknowledging John Abraham's point that the 'regulation of medical technologies, healthcare institutions and pharmaceuticals is far too complex to be encapsulated in legislative acts'.[8] Rather, drawing on particular strands of socio-legal studies that look beyond straightforward 'regulation by the state through the use of legal rules backed by (often criminal) sanctions',[9] I approach regulation as a broad range of customs, rules, guidelines, oversight, 'encompass[ing] anything from codes of practice of professional bodies to traffic lights and signs in a neighbourhood'.[10] Such a flexible approach allows one to engage with the changing features of research ethics review in the UK; for example, the way in which the institutional framework for RECs has changed over time from collegial professional regulation to more formal, governmentally defined structures, or how, even at the same point in time, infringing one's REC approval can result in legal sanctions (if one is carrying out a pharmaceutical trial) or no formal penalty at all (if one is a social scientist interviewing patients). A rigid under-standing of regulation, that sees it either as formal administrative law or social control amongst professionals, fails to engage with the nature of what RECs do and how they reach their decisions.

This introductory chapter starts by setting the origin of RECs in a broad context of developments within the British regulatory state, before introducing the idea that there are anomalies in REC practice that challenge the idea that these bodies are objective and impersonal regulators. Linking to wider studies exploring the role of trust in regulatory decisions, the chapter then engages with methodological issues before setting out the empirical basis for the claims made in this book.

In exploring what kind of thing RECs are, typical accounts of their origin and development in the NHS build on the aftermath of the Second World War and the so-called Nuremburg 'Doctors' trial', sug-gesting steadily increasing and externally imposed oversight and re-striction of research. Tightening of controls might be presented as a response to research 'scandals', emphasising the need for increased control.[11] Similar accounts have been offered of the development of ethics review in the US and IRBs – Institutional Review Boards, the US equivalent of RECs – with the additional insight that changes in such review are driven by academic bioethics, which had begun to expand as a discipline at the same time as IRBs took root. The most obvious link between IRBs and academic bioethics centres on the rules struc-turing IRB decisions – mapped out in the 1978 *Belmont Report* – and the highly influential book, *Principles of Biomedical Ethics*, first published in the same year.[12] As Tom Beauchamp – co-author of *Principles*, and staff member for the Commission which wrote the *Report* – has noted, 'both works were written simultaneously, the one inevitably influencing the other'.[13] While the application of the sets of principles in the two texts diverged somewhat, and the book itself has undergone numerous revisions over the years, academic bioethics clearly played a founding role in IRB decision making.[14]

The same cannot be said of RECs in the UK, where typically the guidance documents for REC decision making originate from the Royal College of Physicians (RCP) in the years running up to the mid-1990s or the Department of Health (DH), more recently. There is little or no evidence of the input of academic bioethics into these documents. The committees involved in writing the RCP's documents were overwhelmingly made up of people from a medical background and the DH guidance was very much 'bottom up', the result of consultation with serving REC members.[15] Given this lack of bioethical input into the development of REC processes it is perhaps unsurprising that academic bioethics is conspicuous by its absence in terms of the actual decision making made in NHS RECs. As Sarah Dyer notes, based on her observation of 20 different Local Research Ethics Committees (LRECs), the idea that these committees apply ethical theory is profoundly mistaken:

> That academic texts and committee meetings are different is hardly a shocking pronouncement. However, it is in this gap between the two that the fallacy of 'applied' rests. Saying that universal principles are taken and applied in practice, in this case by LRECs, obscures far more than it clarifies. LREC discussion does not 'abstract away the particulars' to reveal the universal, rather it reasons through the particulars.[16]

The relationship between academic bioethics and ethics review is one place where there are clear differences between the US and the UK, yet discussions in this area often tend to treat all ethics review bodies (or, more usually, the problems they raise for researchers) as homogeneous, an approach I have, elsewhere, labelled presumed or pseudo-isomorphism.[17] The key point is that, just because, on the surface, IRBs (in the US) and RECs (in the UK) look alike in how they are set up and what they do, does not mean they *are* doing the same thing. Pseudo-isomorphism allows critics of ethics review to draw on a wide range of disparate (even contradictory) examples to attack RECs (or IRBs, or their equivalent) without having to actually present system-specific data. For example, a critic might draw on an example of how IRBs unfairly serve the interests of their host institutions to attack NHS RECs despite the relationship between a hospital and an IRB in the US being quite different from that between an NHS REC and its putative host institution. Thus a key approach in this book is to remain acutely aware of the differences between different ethics review systems, and to draw on data or analysis of non-UK ethics review bodies in a cautious and comparative manner.

Returning to our origin story, if RECs' inception does not lie in academic bioethics, then where do they come from? On the surface, from their origins in the late 1960s through to the current day, the development of RECs in the NHS might be seen as in keeping with broader changes within the British regulatory state. As mapped out by Michael Moran, the trajectory of these changes starts with the Victorian period of 'club government', characterised by its highly cooperative nature where 'resort to sanctions was rare; and there developed an overwhelming stress on fostering trust between regulator and regulated'.[18] This form of 'club regulation', at least as far as medicine and other professions was concerned, was not ignored by the state, but rather acknowledged and devolved, since the:

> state endowed regulatory institutions with authority, but then practised the lightest of light touch controls; the self regulatory institutions themselves, in turn, adopted collegial regulation – a style that presumed control among an elite of equals, was designed to foster collegial solidarity, and relegated hierarchical controls and the exercise of sanctions to a marginal role.[19]

To some extent, this is what we might expect in any area of professionalised practice, and in the UK, it was professional medics – both as individuals and in their representative institutions – who initiated and developed RECs and their practices. Starting in the late 1960s, with individuals' concerns about remaining eligible for US research funding (from the Public Health Service), responsibility for RECs was soon assumed by the RCP of London which, with the support of the Ministry of Health, encouraged the development of RECs across the NHS, and began to set down guidance for their composition and practice.[20]

As explained in Chapter 2, by the late 1980s, the now DH began to take formal responsibility for RECs, issuing its own guidance in 1991 – the so-called 'Red Book' – and setting up a series of Multicentre RECs (MRECs) in 1997 to improve the review of research taking place at multiple sites. From Moran's view of the regulatory state, we might see these changes as part of a broader

> radical shift to formality, including legally backed formality, in regulatory relationships; a shift from tacit to explicit knowledge, in the form of more elaborate codification of rules and more elaborate and onerous reporting requirements; and the reorganization of regulated domains into a reshaped set of hierarchically organized institutions subject to systems of close formal reporting and central surveillance.[21]

It is certainly the case that the end of the 1990s saw an expansion of the DH's rules and guidance documents for RECs, including the *Governance Arrangements for NHS Research Ethics Committees (GAfREC)* document (setting out the broad framework within which RECs were to operate) and the first in a series of standard operating procedures (SOPs) to regularise the way in which applications were processed by committees. In keeping with broader regulatory developments, some of these changes were driven, in part, by international standardisation in ethics committee practice, such as the EU's Clinical Trial Directive (which sets out a range of issues for committees to consider when reviewing a pharmaceutical trial) and the International Conference on Harmonisation's (ICH's) *Guidelines for Good Clinical Practice* (which expand the need for ethics review to pharmaceutical trials).[22]

Yet drawing on Moran, and his use of Michael Power's critique of the 'audit society', one could argue that such changes were just as much the result of a general erosion in trust within British regulatory systems as international developments:

> Power's arguments have a particular bearing on our understanding of regulation in Britain because they cut against the grain of our established understanding of the British regulatory system. A number of landmark studies had painted highly distinctive pictures of British regulation … emphasiz[ing] the importance of relations of trust between regulator and regulated and argu[ing] for the importance of co-operation in effective regulation … Power's book amply demonstrates that it is now passing away. At the root of the audit society is a deficiency of trust in existing regulatory processes and institutions.[23]

From this perspective, as the cosy and, quite frankly, not terribly democratic club government gave way over the twentieth century, so too, albeit with some delay, did the associated club regulation, drawing as it did on professional collegiality and trust. Thus, looking at how RECs have changed over the past 50 years, we might see an example of the 'reconstruction of self-regulatory institutions along more formally organized, more codified, and more state controlled lines [which] was seen everywhere'.[24] As Richard Ashcroft puts it: 'Placing RECs in the governance framework effects the completion of [their] … conceptual transformation …

from peer-led "ethical committees" close in spirit if distant in form to the professional-oriented model, to management and institution-oriented quality management systems.'[25] From this point of view, a result of these changes is that:

> In terms of thinking about the potential harms of research, local and multicentre research ethics committees draw upon a range of ethical frameworks and guidelines … Influenced by public concern and anxiety over medical research, those reviewing research do so through the lens of modern 'risk society', tending to focus on technical assessments of the risk of harm.[26]

In line with these broad changes to regulatory culture, RECs are commonly characterised by medical professionals and other biomedical researchers, as obstructive, overly bureaucratic bodies, requiring an excess of paperwork, over-reaching in terms of the kinds of research they should concern themselves with (and the kinds of features of research they should consider ethically relevant). These complaints take various forms, from analyses of the varying responses from RECs to applications for multisite research,[27] to discussions of proposed, impending, or recent regulatory change.[28] The consistency of concerns over time is striking, with, for example, typical complaints from the mid-1990s including that:

> Administrative matters were handled very poorly in many cases … The issues raised by the committees about the trial cast doubt on their grasp of methodology … having to explain to a local research ethics committee chairman the basic rationale for attempting a randomised trial, rather than performing a retrospective audit, raised concern about their level of scientific expertise.[29]

Six years later, a complainant noted that:

> The bureaucratic unhelpfulness of research ethics committee procedures might be bearable if the committees attended to the ethics of the studies they reviewed. In our experience they have been concentrating on scientific, legal, and confidentiality issues instead of ethical issues.[30]

A further four years on, concern was expressed that 'clinical researchers are exhausted by the demands of ethics committees that seem more concerned with the science (which they cannot necessarily judge) and editorial control of patient information sheets than with ethics'.[31]

While professionals' complaints about RECs' bureaucratic inefficiencies died down following further centralisation post-2010, it remains the case that a commonly held problem of 'prospective research ethics review ... is that it encourages a procedural and legalistic approach to research ethics and fails to capture what is important to ethics'. Indeed:

> the external observer might be forgiven for thinking that the focus is upon process, in the sense that research ethics is about the establishment and refinement of a set of routines for the review of research ... a procedural approach to research ethics review tends to favour a formalisation of the ethical issues, in the sense that research ethics guidelines are increasingly framed as sets of rules.[32]

As a consequence, this perspective presents RECs and IRBs as inimical to trust in researchers, that while 'Investigator virtue is highly valued ... ironically, the compliance culture of modern human-subjects protection assumes that investigators cannot be relied on.'[33] Such a position suggests that, in essence, 'RECs provide a means of separating decisions about the ethics of a study from the people who are involved in designing and running the study – the researchers.'[34]

Yet whatever the *external* impression might be, of SOPs and governance documents and the standardisation and oversight of a system denuded of trust, the *internal* environment of research ethics review – the actual decisions made by RECs as to what research gets to take place and how it should proceed – is suffused with trust.[35] This centrality of trust helps explain behaviours which otherwise would seem rather peculiar. Take, for example, the question of REC access to patient medical records; on a number of occasions, one of the committees I observed – Coastal MREC – made a counter-intuitive decision with regard to the information given to people being recruited into studies. While in theory RECs are allowed to look at the medical records of people recruited into a trial whenever they want, this committee insisted on the removal of the possibility of any such inspection from information given to participants. It is clear that the REC did not want patients told about the committee's right to review medical records, not because such review was not permitted but rather because, in reality, such inspection would not happen. Such oversight would clash with the underlying ethos of ethics review. At the core of the REC process are decisions about whether researchers who are applying for approval can be trusted to do what they say they will do.

Such an idea, that RECs, as regulatory bodies, decide whether researchers can be trusted or not, does not sit well with everyone. For some critics, trusting researchers to do what they say they will do is irresponsible: 'to rely on prospective review of research proposals in the absences of comprehensive monitoring of the projects is to place undue confidence in researchers' intentions rather than their research practices'.[36] For others, the very act of asking researchers to account for their future behaviour is a 'process of institutionalizing distrust', ignoring that 'the efficacy of ethics review in safeguarding morally acceptable research depends on the moral competence and integrity of individual researchers – the very qualities that institutionalized distrust calls into question'.[37] This book steers a middle path between these two points of view, exploring how RECs operationalise decisions about trust in the 'moral competence and integrity of individual researchers', while at the same time avoiding the kind of cynicism that one might expect a 'process of institutionalizing distrust' to result in.

One way of exploring the central role of trust in REC decision making is to examine how RECs deal with applications and researchers *after* they gain approval; for instance, is there any kind of oversight or are there spot checks to make sure that the research is being run in the way the application said it would be? Debates as to how RECs should handle these issues have a long history. In 1974, at an RCP-convened meeting of chairs of ethics committees, the question was raised 'whether an ethical committee had a duty to follow up studies it had approved'. Since in some areas, ethics committees were the same as the more general Research Committee, some did follow up but this was largely accidental, and anyway, 'it was pointed out that the College's Report had recommended that the two should not be the same body'. While some RECs wrote to researchers after a year to ask about progress, and some chairs argued that it 'was important that ethical committees knew what was being carried out under their approval', a strong case was made that 'the research worker had some personal responsibility, the ethical committee was not a "police force"'.[38]

Four years later, the same kind of issues arose, with the Chair of the committee at University College Hospital (UCH) asking whether 'other committees instigated formal follow-ups on research projects. These were not carried out at UCH at present, but at its next meeting the ethical committee would be discussing whether there should be any follow-up, and if so, to what extent'. In response to this, one chair said that his committee 'had considered this, and tried to carry out follow-ups, but had not yet found a satisfactory method'. 'Other chairmen said their committees sometimes followed up research, but that there was a feeling of "spying" on colleagues and a fear that proposals might be driven away.'[39]

Historically, then, it seems clear that a default position of trusting researchers to do what they said they were going to do is a consistent feature of REC practice. More recently, for example, the enquiry into the events surrounding the Continuous Negative Extrathoracic Pressure (CNEP) trial at North Staffordshire Hospital in the 1990s noted of the REC culture of the time:

> the LREC assumed that once they had approved a project then it would be properly carried out as stated ... The LREC were not required to take any steps to ensure that things were working properly ... When questioned about formal procedures for the LREC to monitor researchers in order to ensure that agreed protocols were being implemented, all those interviewed said there was no formal audit. Most referred to the fact that the LREC sent a letter to a researcher about one year after approval, asking for a progress report.[40]

These assumptions about professional trustworthiness extend not just to post-approval 'policing' of research but also to the nature of the application process itself, as the following exchange – part of the General Medical Council's disciplinary process against Andrew Wakefield – highlights: Questioned by the prosecuting barrister: 'Can you explain a little further what you mean by relying on the probity of the doctor?', Stuart Pegg, Chair of the REC that approved Wakefield's research, notes: 'You assumed that someone was filling it out in good faith, and even though you have not asked every tiny little question, would provide you with information that he thought was relevant to your consideration of the application.'[41] While the assumption that RECs should not monitor or audit research after approval may well have started as part of the developing culture of ethics review in the NHS, by the time such activity became formalised, this position had become more explicitly official. In a consultation paper for the 2001 *GAfREC* document that set out the structure and role of NHS RECs, the DH specifically states that a REC 'has no responsibility for subsequent proactive monitoring'.[42]

Given this reluctance to verify post-approval compliance, the question then becomes, what kind of regulator *are* RECs? At a broad level, RECs 'confer legitimacy on the enterprise of medical research and reassure the public-at-large that there are systems of scrutiny',[43] and the core claim of this book is that they do this through assessing and attesting to the competence and trustworthiness of researchers. Unlike the anonymous, impersonal, and ostensibly objective statistical technologies explored for example, by Ted Porter, in his magisterial *Trust in Numbers*, this book emphasises that the kinds of decisions that RECs make do not actually separate decisions about the ethics of a study from the people who are involved in designing and running it.[44] Rather, on the whole, the trust decisions made by RECs *are* personal and face to face rather than anonymous and institutional, and do indeed rely upon personal trusting relationships.

While the focus of this book is on the UK-specific culture surrounding the role of trust in REC decision making, it is worth noting that the features of the UK system that suggest the importance of trust (for example, the lack of post-review monitoring) can also be found in other countries' approaches to ethics review. For example, Charles Bosk and Joel Frader note that, in the US, research ethics review 'is completely forward looking and relies on an honor system: there is rarely, if ever, surveillance to assess compliance'.[45] As the Office of the Inspector General has observed, in the US the IRB review process 'is rooted in trust' and that:

> Although many researchers may presently perceive the IRB review system as bureaucratic, burdensome, and inefficient, they have yet to view the IRB as a watchdog to be feared and altogether avoided. Indeed, IRBs have traditionally approached their reviews in a collegial, non-threatening, non-skeptical manner.[46]

On a broader basis, in his international comparison of ethics review systems, Paul McNeil notes that a:

generalisation about ethics review internationally is that there is very little monitoring of experimentation after its initial approval. Most countries rely on prospective review of research proposals by committees. There are no countries in which monitoring of research is consistently employed.[47]

Of course, it is probably quite likely that the reasons for lack of post-review oversight differ between systems. The normative isomorphism of the UK system – rooted as it is in professional practice and self-regulation – will probably contrast significantly with the reasoning underlying the design of far more recently founded ethics review systems. Yet even in those countries where ethics committee origins lie in the realms of coercive isomorphism and the need to conform to international standards around research governance, there is considerable reluctance to allow RECs to review research once it has been approved.[48] A good example of this comes from Rachel Douglas-Jones' exploration of the training courses run by international providers for RECs in developing countries, where a Dr Dipika, Chair of an Indian ethics committee:

> was presenting a new scheme that her committee had pioneered ... requiring dedicated members from the committee to familiarise themselves with the documents for a trial, go to the site, meet the Principal Investigator, observe consent processes, look at consent forms, and check thoroughly that everything was in order. She listed 'violations' that her monitors had found as a result of this new procedure, and the assembled audience – most of whom were committee members themselves – knew these findings were serious ... But the discussion after her talk took an entirely different turn: we talked about whether she and her committee should have been monitoring in the first place.

For the organisation training these chairs, the issue was that the pharmaceutical company sponsoring the research had its own clinical monitors, and it was they, rather than the REC, who should oversee the research. Reflecting on this chair's innovation, Douglas-Jones articulates Dr Dipika's reasoning:

> She had doubted whether researchers were able to follow the protocol they had submitted, and whether they respected the decision of the committee and the changes it had requested. She had found in her investigations that they did not. Had she not proven

that doubt was justified? Now that she had shown that researchers could not be trusted to do what they had promised the committee, was greater vigilance not required? Surely her initiative improved the system, made her committee perform better, and increased both the trustworthiness of the data coming out of the trials and, by extension, the committee itself? The opposite seemed to be true. Her new system meant that the committee had overstepped their role within another system: the international system in which FERCAP [the training organisation] was working to include Asian Ethics Review committees.[49]

While the reason why RECs avoid monitoring of research may differ between national systems – historical roots in professional regulation in one system, concerns about interfering in other groups' remit in another – the end result, the importance of trust decisions, may remain the same. The rest of this chapter explores in more detail the reasons that have placed trust at the core of REC decision making before setting out the approach adopted in this book.

If Michael Moran's outline of change in the British regulatory state is correct, then the kind of regulation performed by RECs is something of an anomaly. While the broad direction of change has been away from 'club regulation' towards more impersonal oversight based on monitoring and audit, RECs – in their daily practice at least, if not the broader structures within which they reside – have seemingly resisted this, retaining trust decisions at the core of what they do, and avoiding changes (such as increased post-approval monitoring) which infringe on this. As a starting position, it is important to note that, while some elements of this resistance lie in the strength of the medical profession, this cannot be the whole answer, in part because in *other* areas of their practice, for example clinical treatment of patients, doctors' self-regulation in the UK has been rolled back and standardised in a way very much in keeping with Moran's ideas.[50]

Thus, thinking more deeply about the nature of regulation should lead us to consider the possibility that trust decisions are important, not just in a strange, outlier regulatory practice like ethics review, but in regulatory decisions more broadly. As a number of empirical studies have pointed out, interpersonal trust between regulators and those they oversee can play a crucial role in regulatory decision making. Studies such as Keith Hawkins' classic analysis of health and safety inspection – *Law as Last Resort* – highlight how 'a continuing or personal relationship … between inspector and employer … creates the conditions for the growth of trust' and hence the 'compliance strategy' that Hawkins suggests characterises this kind of regulation (and which is typified by low levels of prosecution).[51] At the core of the insight that regulation almost always necessitates some form of trust is the pragmatic challenge that 'Regulators cannot supervise every regulated activity 24/7, so regulation cannot provide 100 per cent certainty; it encompasses irreducible uncertainty and risk.'[52] As considerable scholarship in organisational studies has explored, trust-based decisions are extremely resource-efficient.[53]

In addition to being a form of regulation (and hence requiring some form of trust decision), REC review is anticipatory, taking place before the fact, before the activity being regulated actually occurs, and thus opens even more space up to the realm of trust. Of course, there are other ways in which biomedical research could be regulated, many of which are after the fact – prosecution of wrongdoing, for example. But given that ethics review is *prior* review, then it is hardly surprising that, unlike regulation which reviews activities that have already taken place, the additional uncertainty of trying to decide researchers' future behaviour requires RECs to make decisions about applicants' trustworthiness.

As I have suggested elsewhere, the nature of the trust relationship in REC decisions is characterised by a double asymmetry (typical of many regulatory relationships).[54] Firstly, unlike many other empirical case studies of trust decisions, where the parties involved are on a level footing and can opt out of an exchange if they do not trust the other party, researchers have to gain RECs' trust, *whether or not they themselves trust the REC and its members*. That the literature, replete as it is with researchers' complaints about the inadequacies of research ethics review, suggests that there is very little trust in RECs (or IRBs, or their jurisdiction-specific equivalent) makes no difference to the requirement that the same researchers, if they wish to remain within the boundaries of regulated research, have to submit to the very same RECs (or IRBs, or their jurisdiction-specific equivalent). Lack of trust in a regulator does not exempt one from having to seek regulatory approval. While in broader terms it may be important that society as a whole should trust regulators to do their jobs, in the case of the day-to-day activities of regulatory bodies, trust decisions are one-way.[55]

The second asymmetry that characterises REC trust decisions (and regulatory decisions more widely) is that the harms of a poor trust decision rarely, if ever, impact directly on the REC itself. The person who bears the cost of the decision to approve an application from an inept or dishonest researcher is not the REC, but the patient enrolled in their research. Such an asymmetry is a challenge for many theoretical accounts of trust which 'tend to be premised on the general idea that actors become, in some ways, vulnerable to one another as they interact in social situations',[56] that, in essence, 'To trust is to act in such a way as to give another agent power over us.'[57] Yet when a REC decides to trust an applicant and approve their research it has not, in any meaningful sense, become 'vulnerable'. Thus, the second asymmetry is that RECs make trust decisions *on behalf of* research subjects. If trust is about risk, it is not RECs that are exposed to the risks about which they make trust decisions.

This book, therefore, sets out to explore this atypical kind of trust, as a way of thinking about what it is that RECs actually do – how they regulate modern biomedical research. In adopting an empirically minded approach to research ethics review, this book follows a 'path less trodden' by social scientists, whose interactions with RECs (and other ethics review bodies) have tended to be characterised by bad-tempered criticism and barely contained outrage, with prior review of social scientific research by IRBs in the US being branded 'unconstitutional' or a form of censorship.[58] Meanwhile, in the UK, senior sociologists such as Robert Dingwall worry that 'Participant-observation may be dying at the hands of philistine IRBs', arguing that such review is itself unethical and that – echoing Margaret Thatcher's call regarding terrorists and publicity – researchers 'need to deprive these bodies of the oxygen of legitimacy'.[59] Similarly, Martin Hammersley, another senior social scientist, rails against the 'evils' of ethics review, concerned that it is 'strangling' research.[60]

However, in parallel with, though largely overlooked by, these social scientific complaints about ethics review is a longstanding body of research into (first) IRBs and (later) NHS RECs, which seek to empirically understand the nature of ethics review systems and processes. And it is to this literature that the present book aims to contribute. Historically these studies date back to the early 1970s, with a majority of those offering insight into internal IRB processes or decision making drawing on self-reportingr via surveys of members of IRBs, an approach echoed in subsequent studies of NHS RECs.[61]

Methodologically (and problematically), a key feature of these studies, and more qualitative interview-based work, is the way in which respondents' self-reporting is seen as providing direct insight into the decision-making process of the IRB (or REC) being studied. Typically, these studies assume a straightforward relationship between what an informant (more often than not, the Chair of a committee) claims an IRB (or REC) does, and how it actually behaves. For example, in his book-length interview-based study, Robert Klitzman argues that 'a great many essential questions remain concerning how IRBs in fact operate, make decisions about specific protocols, and view and fill their roles', going on to suggest that, based on his interviews,

> IRBs seek to work by consensus, discussing issues and having members who wish to express an opinion about a certain study do so ... On most issues, IRBs reach agreement fairly readily ... Yet for a significant number of other studies, achieving uniformity can be hard.[62]

How such confident conclusions about the internal behaviours of IRBs can be drawn on the basis of interviews and a survey is hard to say and, of course, sociologists have pointed out for many decades that such a straightforward interpretation of non-observed behaviour, simply on the basis of a respondent's claims, is deeply flawed. As one of the earliest ethnographic investigations of research ethics review noted, such studies 'lack observationally grounded data on the complicated business of decision making in the complicated world of research ethics'.[63]

In broader sociological debates, at least as long ago as 1957, Howard Becker and Blanche Geer noted the influence of what they called 'Distorting Lenses' on interview data, highlighting how, 'In many of the social relationships we observe, the parties to the relation will have differing ideas as to what ought to go on in it, and frequently as to what does in fact go on in it.' Indeed, they suggest that 'such mythology will distort people's view of events to such a degree that they will report as fact things which have not occurred, but which seem to them to have occurred'.[64]

Colin Jerolmack and Shamus Khan have echoed this, pointing out that 'Not only is it the case that people commonly act in ways that are inconsistent with their expressed attitudes, they also routinely provide inaccurate accounts of their past activities', going on to argue that 'privilege[ing] verbal accounts ... to explain social behaviour ... and making little effort to validate the assumption that the attitudes ... [being] ... measure[d] are associated with the behaviors' in question is to 'commit what we call the attitudinal fallacy – the error of inferring situated behavior from verbal accounts'.[65] And, of course, in the context of researching ethics review, it is a considerable leap to go from the perfectly reasonable claim that the 'perspectives of IRBs themselves are essential' to the assumption that these perspectives (furnished by interviews with IRB members) must necessarily provide a direct and accurate representation of the decision-making practices of these bodies.[66]

More obviously, beyond the fundamental attitudinal fallacy that characterises so many of these studies, there is an underacknowledged issue around the *kinds* of IRB or REC member who tend to end up getting interviewed or filling out a questionnaire – predominantly the chairs of, or the administrators serving, these committees.[67] A single informant – a powerful male medical chair, for example – may present his committee as wholly consensual, while other members of the same IRB or REC – lay members, community representatives, female nurse members, for example – may have a very different view on the character of its decision making. It is not that it is not interesting to interview chairs of IRBs or RECs about how they think their committees operate, but rather that the opinion of these very specific individuals should not be seen as providing a direct representation of how that body behaves.

Urging caution about what interview or questionnaire studies can tell us about research ethics review is not, of course, the same as saying they can tell us nothing. Indeed, it is clear that there are studies based solely on external-facing information – i.e. not even including interviews – that are extremely informative. We might think of Richard Nicholson, using only the annual reports that RECs are required to produce, showing how much insight (into variation in composition and practice, for example) such putatively dry, 'external' data can provide.[68] More recently, Mary Dixon-Woods and colleagues' analysis of the letters that NHS RECs use to communicate with applicants has provided important evidence as to the ways in which such bodies publicly justify their decisions. Yet, as Clapp and colleagues note regarding this kind of research, however important such letters are in the public justification of such decisions 'by the time ... [they] ... arrive in the mailbox of researchers, they bear little trace of the social relations that produced them'.[69] Studying the social relations that produce such decisions requires observation of these committees in a naturalistic setting. Coming to understand not just who shapes the final decisions made by such bodies, or how applications come to be approved or rejected, but more complex issues such as the role of trust in shaping these processes means we need to study REC decision making 'in the wild'. Similar methodological conclusions come from consideration of Science and Technology Studies (STS) and its long history of enquiry around regulation, focusing on the characteristics and importance of 'regulatory science', a particular form of enquiry that feeds into policy making, legal decisions, and regulatory practice, and related concepts such as 'regulatory objectivity'.[70] A common approach offered in such work is to provide comparative analysis – either over time or between jurisdictions – of specific case studies or regulatory systems more broadly to highlight the variation that exists in science-based regulation.[71] Methodologically, these studies have tended to employ retrospective analysis – for example using documents and interviews with key actors – to reconstruct individual regulatory decisions and to explore the social, cultural, and institutional factors underpinning them.[72] Such approaches make sense given the normally restricted nature of regulatory organisations, but also mean that there is clear need for methodologically complementary studies that draw on observational data and which examine such decision making in a more 'real-time' fashion.[73]

While such ethnographic approaches to research ethics review are rarer than interview- or questionnaire-based studies, scholarship in this area, most obviously Laura Stark's work on US IRBs, highlights the way in which individual review bodies develop their own standards and approaches to ethics, as well as the way in which IRBs draw on applications and other documents to shape their decisions.[74] Other studies have explored how RECs in the UK draw on written texts to shape their decisions; how equivalent bodies in the Netherlands draw on different evaluative repertoires;

how institutional context serves to structure committee decision
making; or how the idea of 'reasonableness' is applied in Australian
ethics committees.[75] More recently, Edward Dove has examined
NHS REC decision making in the context of regulatory changes
over the past decade.[76]

Building on an earlier insight arising out of my own work, some
of these studies have begun to explore the importance of trust in
research ethics review, but there remain considerable gaps in our
ethnographic understanding of the interactional aspects of REC
decision making.[77] This book presents just such an ethnographic
study, drawing on different sets of data with the core of my analysis based on observations carried out between July 2005 and
November 2006 at three UK RECs: St Swithin's, Northmoor, and
Coastal.[78] St Swithin's LREC was founded in the late 1960s at the
teaching hospital after which it is named and which is internationally renowned for its biomedical research. Despite no longer having
formal links with the hospital, the REC draws almost all its expert members and most of its applicants from St Swithin's or one
of its satellite institutions. Northmoor and District LREC arose
from the merger of two other committees in and around the town
of Northmoor which took place about a year before my observations began. The committee reviews a mixture of applications from
local institutions and from around the country. Finally, as one of
the original ten MRECs set up in 1997, Coastal MREC has never
had any institutional affiliation, drawing its members from a wider
geographical area than the LRECs, with many members previously
having served on local committees. As an MREC, of my three committees, Coastal received the majority of pharmaceutical clinical
trials that I observed.

These committees are not statistically representative of NHS
RECs but they do display a number of characteristics that mean
they can be considered 'typical'.[79] They reviewed a wide range of
studies, from pharmaceutical-backed clinical trials (phases II–IV),
through surgical trials, public health research, qualitative social science, and student (postgraduate and undergraduate) research. They
all conformed to the DH rules regarding the composition and processes of such committees.

The sampling strategy offered here – long-term engagement with
a small number of sites – contrasts with a number of other putatively

ethnographic studies of research ethics review bodies which have
sought to explore individual meetings at a wide range of different
committees, either on a national or international basis.[80] While
such studies generate data from many more sites than the (rather
old-fashioned) approach offered in this book, they also have considerable disadvantages if one is interested in understanding, in an
in-depth way, how ethics review bodies make decisions. A number
of problems arise with only one set of observations from each committee (i.e. a single meeting). The first is that, on occasion, rare
events can take place in a single meeting, but without comparison
with other meetings *of the same committee*, the ethnographer is
unable to say how typical such events are for that particular group.
Without exposure to the specific practices of that specific committee over time, one cannot really say whether the rare event in
question is, indeed, rare.[81] The second, related, problem centres
on the nature of initial observations in a new research site (which
each meeting is for these researchers) and how unfamiliarity with
each setting generates observational artefacts. The most obvious
example of this is a particular pattern of decision making set out
by Fitzgerald, where she claims REC members catch themselves in
a moment of inattention, and the resulting refocusing – what she
refers to as a form of 'moral panic' – results in stricter decision
making for the next application.[82] However, in my own study, attempts to replicate this observation failed; while, of course, REC
members' attention varied in a patterned way (e.g. just before a
break after a long session), this was not, in my data, associated
with a subsequent stricter application of the rules. It seems to me
that the most likely explanation is that Fitzgerald's moral panics
were artefacts of the initial observation at each site, a result of the
observer being unfamiliar with the members and specifics of each
committee when data was gathered.

While there has been recent interest in empirical examination of
the efficiency and effectiveness of IRB and REC decisions, this is not
the aim of this project. The ethnographic approach adopted in this
book does not allow one to assess the validity of REC decisions,
whether the things that committee members *think* make applicants
trustworthy *actually* mean that these researchers are less likely to
infringe their ethics approval. Rather the aim is to explore the internal ways in which RECs reach their decisions about applicants,

regardless of whether or not these decisions make sense to people outside these committees. From this perspective, one of the key empirical questions is, how do those making such decisions decide whether someone is trustworthy or not? The solution is to draw on work that 'treat[s] trust as the product of underlying trustworthy-making character features'.[83] Such an approach focuses on those characteristics of an interaction which lead one party in a relationship to conclude that the other is trustworthy.

In terms of motivation, this approach avoids a narrow model – such as 'encapsulated interest' theory – proposing rather: 'a wide variety of sources of trustworthiness, which include not only one's self-interest but also moral principles, social norms, and even specific dispositions that … can make one trustworthy'.[84] A central challenge in deciding whether someone is trustworthy then becomes working out whether they possess these 'sources of trustworthiness', which Bacharach and Gambetta call 'trust-warranting properties'. Many of these properties, aspects of one's character for example, or one's values, are not directly observable; rather a truster has to depend upon a range of external 'signs' which indicate the presence of such. For example, one might observe 'physiognomic features – the set of the eyes, a firm chin – and behavioral features – a steady look, relaxed shoulders – and treat … them as evidence of an internal disposition'.[85] In the specific example of research ethics review, if a core aspect of what RECs do is make trust decisions about applicants and applications, then these decisions must revolve around the detection and interpretation of applicants' trust-warranting properties, and the signs which indicate their presence.

The adoption of this perspective does not necessarily commit oneself to the kind of 'cognitive' theory of trust often associated with this approach, and which has been the subject of some criticism.[86] Rather, focusing on trust-warranting properties and their external signs is a solution to the practical empirical problem of how to document how trust decisions get made, whether the processes underpinning those decisions are purely cognitive, emotional, or some combination.

To return to my sources of data, alongside observations between 2005 and 2008 I carried out semi-structured interviews with 31 members of these three committees (out of a possible 48) as well as administrators or 'coordinators' for two of them. I interviewed members of six other RECs identified because of their role (lay chair of a REC), the committee that they sat on (a specialist children's hospital), or their long experience of sitting on RECs (giving insight into changes over time). Additional material was gathered from interviews with eight policy makers, six representatives of pharmaceutical companies, and six commentators – academics or medics who have written about RECs. All interviews use pseudonyms and were recorded, transcribed, and analysed with the help of Nvivo QDA software. Observations and interviews, and their subsequent analysis, were not focused on any one aspect of REC practice, but rather sought to provide a detailed description of what goes on in a meeting.

The third set of data for this study is archival and comes from material related to research ethics review, medical experimentation, and related topics taken from the UK National Archives (NA) covering a 20-year period from the late 1960s onwards. Documents in this archive include minutes from meetings, letters, memos, and reports circulated among the main institutional policy players in this area – the DH and its predecessors, Medical Research Council (MRC) and the RCP. This material is supplemented by documents from the archives of the RCP, including minutes from meetings of an informal network, and the chairs of RECs, which met from 1974 to the late 1980s. Material on the Chief Medical Officer's Consultative Group on Research Ethics, which met in the mid-1990s, was available from the personal archive of Professor Naomi Pfeffer. My aim in drawing on this material was to explore the 'depth' of particular aspects of research ethics review. If we start from a premise that REC decision making is, at least in part, a culturally determined practice shaped not just by the job that RECs perform (i.e. decide whether a particular piece of research should go ahead) but their context (an NHS that is continually restructuring yet remains with a central core goal) and membership (experts and lay members from a range of different backgrounds), then we might suspect that particular debates and ways of doing things might have an extensive institutional history. This data, therefore, sheds light on the historical depth certain practices and beliefs have within the REC community.

Finally, additional insight into REC decision making came from the events surrounding the clinical trial of a drug called TGN1412, which took place in the middle of my fieldwork. On the morning of 13 March 2006, eight young men woke up in a building next to

Northwick Park Hospital, in North London. At 8 a.m., after a series of medical checks and procedures these healthy volunteers were, in turn, injected with an experimental drug labeled TGN1412, a proposed treatment for B-cell chronic lymphocytic leukaemia and rheumatoid arthritis. Setting up and injecting the drug into their arms took around 10 minutes, so it was while the final man was being treated (just after 9 a.m.) that the first volunteer, David Oakley, complained of a headache. Over the next few hours, the six men who had received the drug (as opposed to the two recipients of placebo) became extremely distressed and experienced a series of symptoms, including nausea, vomiting, back pain, increased body temperature, and drops in blood pressure. Staff in the building, the clinical trials unit of a private company called Parexel, became so concerned that by midnight, 16 hours after the original injections, all six volunteers had been moved to the intensive care unit of the hospital next door. The drug had triggered a 'cytokine storm', a massive over-stimulus of the immune system which left the men gravely ill and hospitalised for several weeks. The ensuing press coverage ensured that this 'adverse event' (to borrow the terminology of clinical trials) – dubbed the 'elephant man trial' because of how the head of one of the volunteers swelled up – became world-famous. Eventually, all six volunteers left hospital, although one of them, Ryan Wilson, has had all his toes amputated as a result of a form of gangrene, and long-term effects (such as cancers and immune-related illnesses) have been predicted.[87]

Unsurprisingly, these events had a noticeable effect on the RECs I was observing. While much of the press coverage focused on the failure of the medical regulator (the Medicines and Healthcare products Regulatory Agency) to spot the possible risks of the short-interval dosing scheme being used, the committees were acutely aware that the trial had been approved by a fellow REC and that, in different circumstances, it could have been their committee under scrutiny.[88] Using a freedom of information request and public documents, I identified and subsequently interviewed members of the REC involved (the Brent REC)[89] as well as people involved in running the trial, with this data providing an opportunistic way of testing the claims of scholars such as Diane Vaughan that such organisational disasters tend to result not from the *failure* of regulatory processes, but rather the typical application of regulatory procedures that, on many other occasions, have produced entirely benign results.[90]

The next four chapters of this book draw on this empirical data to explore how RECs make trust decisions by examining various kinds of information, including: the written material researchers submit (Chapter 1); local information about applicants arising as a result of repeat applications or working alongside researchers in a clinical setting (Chapter 2); and the 'performance' applicants provide in face-to-face meetings with RECs (Chapter 3). Chapter 4 provides a final empirical discussion, focusing on the way in which the longstanding debates around whether RECs should involve themselves in *scientific* review of applications are rooted in concerns around trust decisions. The final chapter attempts to explore the wider implications, both for the regulation of biomedical research and regulatory studies as a whole, of the central role of trust in research ethics review.

Notes

1 Steven Shapin, 'Trust, Honesty, and the Authority of Science', in Ruth Ellen Bulger, Elizabeth Meyer Bobby, and Harvey V. Fineberg (eds.), *Society's Choices: Social and Ethical Decision Making in Biomedicine* (Washington, DC: National Academy Press, 1995), 396.
2 John Hardwig, 'The role of trust in knowledge', *The Journal of Philosophy*, 88:12 (1991), 706.
3 Clare Dyer, 'Wakefield was dishonest and irresponsible over MMR research, says GMC', *British Medical Journal* 340 (2010), c593; Zosia Kmietowicz, 'Wakefield is struck off for the "serious and wide-ranging findings against him"', *British Medical Journal*, 340 (2010), c2803.
  The formal legal requirement for REC approval of pharmaceutical trials dates from 2004 and the enactment of the EU Clinical Trials Directive into UK law: The Medicines for Human Use (Clinical Trials) Regulations 2004/1031. In terms of professional misconduct, doctors have been required to have REC approval to share patient data since 1977 (General Medical Council, *Professional Conduct and Discipline* (May 1977), 16) and, more generally, to 'Check that the research protocol has been approved by a properly constituted research ethics committee' since 1995 (General Medical Council, *Good Medical Practice* (October 1995), 13). Between 1991 and 2018 the General Medical Council disciplined doctors on at least eight occasions for failing to gain REC approval for research or failing to abide by the letter of such approvals where given.
4 Steven Shapin, *The Scientific Life: A Moral History of a Late Modern Vocation* (Chicago, IL: Chicago University Press, 2008), 3.

5   Steven Shapin, 'Trust, Honesty, and the Authority of Science', 402.

6   Steven Shapin, *A Social History of Truth: Civility and Science in Seventeenth Century England* (Chicago, IL: The University of Chicago Press, 1994), 7.

7   Steven Shapin, *The Scientific Life*, 4–5.

8   John Abraham, 'The politics and bio-ethics of regulatory trust: Case-studies of pharmaceuticals', *Medicine, Health Care and Philosophy*, 11 (2008), 415.

9   Julia Black, 'Critical reflections on regulation', *Australian Journal of Legal Philosophy*, 27 (2002), 2.

10   Edward Dove, *Regulatory Stewardship of Health Research: Navigating Participant Protection and Research Promotion* (Cheltenham: Edward Elgar Publishing, 2020), 3.

11   For the idea that these regulatory developments are a form of 'moral panic' in response to a series of research scandals, see: Phillip Pettit, 'Instituting a research ethic: Chilling and cautionary tales', *Bioethics* 6:2 (1992), 90–112; D. Chalmers and Phillip Pettit, 'Towards a consensual culture in the ethical review of research', *Medical Journal of Australia* 168:2 (1998), 79–82; Barry Bozeman and Paul Hirsch, 'Science ethics as a bureaucratic problem: IRBs, rules, and failures of control', *Policy Sciences* 38:4 (2006), 269–291. I have argued elsewhere that such an explanatory framework has little relevance to the development of NHS RECs: Adam Hedgecoe, 'Scandals, ethics and regulatory change in biomedical research', *Science, Technology, & Human Values*, 42:4 (2017), 577–599.

12   National Commission for the Protection of Human Subjects of Biomedical and Behavioral Research, Department of Health, Education and Welfare (DHEW), *The Belmont Report* (Washington, DC: United States Government Printing Office, 1978); Tom Beauchamp and James Childress, *Principles of Biomedical Ethics* (New York: Oxford University Press, 1978).

13   Tom Beauchamp, 'The Origins, Goals, and Core Commitments of The Belmont Report and Principles of Biomedical Ethics', in Jennifer Walter and Eran Klein (eds.), *The Story of Bioethics: From Seminal Works to Contemporary Explorations* (Washington, DC: Georgetown University Press, 2003), 17.

14   For more detail on this see: Renée Fox and Judith Swazey, *Observing Bioethics* (New York: Oxford University Press, 2008), 128–142.

15   The Committee on the Supervision of the Ethics of Clinical Investigations in Institutions, which made the original recommendations to set up RECs in 1967, consisted of ten members, all doctors; and while the details of the first committee to draft the RCP's formal Guidelines for RECs in 1984 are unclear, the committee that drafted

the second edition (1990) had 23 members, including six non-medical members; and the committee for the third (1996) edition also consisted of 23 members, of whom only five were non-medical lay members. Interviews with policy makers involved in drafting NHS policy in this area, for example the so-called *GAfREC* document of 2001, made clear that this process was shaped by input from REC members rather than academic bioethicists.

16 Sarah Dyer, 'Applying Bioethics: Local Research Ethics Committees and their Ethical Regulation of Medical Research' (PhD thesis, University of London, 2005), 74. Indeed, we might follow Dyer in reasoning that the rejection of academic bioethics in REC practice is not unrelated to the nature of RECs as a form of professional self-regulation: 'committee members are therefore quite unwelcoming of academic or professional Bioethics, which they construe as external standards that threaten their professional autonomy. As a form of regulation, this is essentially self-regulation of medical professionals by medical professionals': ibid., 110.

This does not mean, of course, that academic bioethicists in the UK do not make claims about the importance of their discipline to REC decision making – 'Proper ethics review requires a sound philosophical approach, and committees need an understanding of the ethical principles and concepts that underpin research ethics review' (J. Savulescu, 'Two deaths and two lessons: Is it time to review the structure and function of research ethics committees?' *Journal of Medical Ethics*, 28:1 (2002), 1–2) – only that there is little evidence of their success at influencing REC practice.

17 Adam Hedgecoe, 'The problems of presumed isomorphism and the ethics review of social science: A response to Schrag', *Research Ethics* 8 (2012), 79–86.

18 Michael Moran, *The British Regulatory State: High Modernism and Hyper-Innovation* (Oxford: Oxford University Press, 2003), 42.

19 Michael Moran, *The British Regulatory State*, 51. The similarities between Moran's comments on Victorian industrialists and Shapin's analysis of Restoration scientists is striking: 'Observance of rules was assumed to be the normal state of affairs since the subjects of regulation were gentlemen who could be trusted; most breaches of rules were thus viewed as mere formal or technical irregularities; and sanctions were reserved for a deviant minority.' Michael Moran, *The British Regulatory State*, 47.

20 For more of the early history of RECs in the UK, see Adam Hedgecoe, ' "A form of Practical Machinery": The origins of Research Ethics Committees in the UK: 1967–1972', *Medical History*, 53:3 (2009), 331–350; Jenny Hazelgrove, 'The old faith and the new science: the

Nuremberg Code and human experimentation ethics in Britain,
1946–73', *Social History of Medicine*, 15:1 (2002), 109–135; Jenny
Hazelgrove, 'British research ethics after the second world war: the
controversy at the British Postgraduate Medical School, Hammersmith
Hospital', in Volker Roelcke and Giovanni Maio (eds.), *Twentieth
Century Ethics of Human Subjects Research* (Stuttgart: Franz Steiner,
2004), 181–197.
    The RCP Guidance is offered in a series of documents: *Guidelines on
the Practice of Ethics Committees in Medical Research* (London: Royal
College of Physicians of London, 1984); *Guidelines on the Practice
of Ethics Committees in Medical Research Involving Human Subjects*
(London: Royal College of Physicians of London, 1990, 2nd edition);
*Guidelines on the Practice of Ethics Committees in Medical Research
Involving Human Subjects* (London: Royal College of Physicians of
London, 1996, 3rd edition).

21 Moran, *The British Regulatory State*, 79–80.
22 Department of Health, *Governance Arrangements for NHS Research
Ethics Committees* [*GAfREC*] (London: Department of Health,
2001); COREC, *Standard Operating Procedures for Research Ethics
Committees*, Version 1, February 2004.
    While the Clinical Trials Directive – Directive 2001/20/EC – only
came into effect in 2004, from talking to UK policy makers it is
clear that it influenced the REC system – for example, the *GAfREC*
document – from 2001 onwards. The International Conference on
Harmonisation (ICH) attempts to standardise drug regulation and approval processes between the US, UK, and Japan. Its guidelines on the
running of clinical trials – International Conference on Harmonisation,
Guidelines for Good Clinical Practice, 1996 – makes ethics committee
review a necessity for pharmaceutical trials.
23 Michael Moran, 'Understanding the regulatory state', *British Journal
of Political Science*, 32:2 (2002), 406–407.
24 Moran, *The British Regulatory State*, 92.
25 Richard Ashcroft, 'The ethics and governance of medical research: What
does regulation have to do with morality?', *New Review of Bioethics*,
1:1 (2003), 52–53. For a detailed attempt to describe the development
of medical ethics in the UK (including the early years of RECs) in terms
of Moran's concept of 'club regulation', see Chapter 1 of Duncan
Wilson, *The Making of British Bioethics* (Manchester: Manchester
University Press, 2014).
26 Sara Shaw and Geraldine Barrett, 'Research governance: Regulating
risk and reducing harm?' *Journal of the Royal Society of Medicine*,
99:1 (2006), 15 and 19.

27 Tamzin Berry, Tony Ades, and Catherine Peckham, 'Too many ethical committees', *British Medical Journal*, 301 (1990), 1274; R. Benster and A. Pollock, 'Ethics and multicentre research projects', *British Medical Journal*, 304:6843 (1992), 1696; P. Garfield, 'Cross district comparison of applications to research ethics committees', *British Medical Journal*, 311:7006 (1995), 660–661; Joanna Tully, Nelly Ninis, Robert Booy, and Russell Viner, 'The new system of review by multicentre research ethics committees: Prospective study', *British Medical Journal*, 320:7243 (2000), 1179–1182; David Wald, 'Bureaucracy of ethics applications', *British Medical Journal*, 329 (2004), 282–284; Paul Stewart, Anna Stears, Jeremy Tomlinson, and Morris Brown, 'Regulation – the real threat to clinical research', *British Medical Journal*, 337 (2008), a1732.

28 T. Marshall and P. Moodie, 'Research ethics committees revisited', *British Medical Journal*, 299:6713 (1989), 1419–1420; S. Lock, 'Monitoring research ethical committees', *British Medical Journal*, 300:6717 (1990), 61–62; P. Moodie and T. Marshall, 'Guidelines for local research ethics committees', *British Medical Journal*, 304:6837 (1992), 1293–1295; K.G. Alberti, 'Multicentre research ethics committees: Has the cure been worse than the disease?', *British Medical Journal*, 320:7243 (2000), 1157–1158; R. Nicholson, 'Another threat to research in the United Kingdom', *British Medical Journal*, 328:7450 (2004), 1212–1213.

29 Z.J. Penn and P.J. Steer, 'Commentary: Local research ethics committees: Hindrance or help?', British *Journal of Obstetrics and Gynaecology*, 102 (1994), 1–2.

30 Jon Nicholl, 'The ethics of research ethics committees', *British Medical Journal*, 320 (2000), 1217.

31 Charles Warlow, 'Clinical research is under the cosh again', *British Medical Journal* 329 (2004), 241–242.

32 Angus Dawson, Sapfo Lignou, Chesmal Siriwardhana, and Dónal P. O'Mathúna, 'Why research ethics should add retrospective review', *BMC Medical Ethics*, 20: article 68 (2019), 2.

33 D.S. Jones, C. Grady, and S.E. Lederer, 'Ethics and clinical research – the 50th anniversary of Beecher's bombshell', *New England Journal of Medicine*, 374 (2016), 2393–2398, 2397.

34 Mary Dixon-Woods and Richard Ashcroft, 'Regulation and the social licence for medical research', *Medicine, Health Care and Philosophy*, 11:4 (2008), 388.

35 A similar contrast can be found in Michelle Pautz's work on environmental regulation, where she contrasts 'the overwhelmingly positive nature of the interactions between inspectors and facility personnel'

with a 'regulatory system that is often thought to be adversarial'. She suggests that such 'adversarialism is more endemic during the creation of regulations rather than on the frontlines of implementation': 'Trust between regulators and the regulated: A case study of environmental inspectors and facility personnel in Virginia', *Politics and Policy*, 37:5 (2009), 1065.

36 Paul McNeil, *The Ethics and Politics of Human Experimentation* (Cambridge: Cambridge University Press, 1993), 111.

37 Linus Johnsson, Stefan Eriksson, Gert Helgesson, and Mats G. Hansson, 'Making researchers moral: Why trustworthiness requires more than ethics guidelines and review', *Research Ethics,* 10:1 (2014), 29.

38 Royal College of Physicians Archive, London (hereafter RCPA), MS4930, 'Report of a meeting of the Chairmen of Ethical Committees or their representatives held on Thursday, 17 October 1974', 8.

39 RCPA, MS4930, 'Report of a meeting of the Chairmen of Ethical Committees or their representatives held on Tuesday, 26 September 1978', 16. See also: Emma Pickworth, 'Should local research ethics committees monitor research they have approved?' *Journal of Medical Ethics* 26:1 (2000), 330–333.

   For a US-focused discussion of unease at IRB monitoring of research post-approval, see: Richard Saver, 'Medical research oversight from the corporate governance perspective: Comparing Institutional Review Boards and corporate boards', *William & Mary Law Review*, 46:2 (2004), 693.

40 NHS Executive West Midlands Regional Office, *Report of a Review of the Research Framework in North Staffordshire Hospital NHS Trust* (Griffiths report) (Leeds: NHS Executive, 2000), 21 and 40.

   The Continuous Negative Extrathoracic Pressure or CNEP trial ran between 1989 and 1993 in Stoke on Trent in Staffordshire. It attempted to discover whether newborn babies with breathing difficulties did better in incubators with lower than normal air pressure. Following complaints from parents that they were unaware their children had been enrolled in experimental treatment, a number of formal enquiries took place. See: Judy Jones, 'Government sets up inquiry into ventilation trial', *British Medical Journal,* 318 (1999), 533; Edmund Hey and Ian Chalmers, 'Investigating allegations of research misconduct: The vital need for due process', *British Medical Journal*, 321 (2000), 752–755.

41 General Medical Council, Fitness to Practise Panel (Misconduct), Case of Andrew Wakefield, John Walker-Smith, and Simon Murch, Transcripts, Day 9, 26 July 2007, 62. For more information on this case, see endnote 3 above.

42 Department of Health, *Governance Arrangements for NHS Research Ethics Committees – Draft Consultation Paper* (London: The Stationery Office, London, 2001), para. 7.16, 13, quoted in: Paul Ramcharan and John R. Cutcliffe, 'Judging the ethics of qualitative research: Considering the "ethics as process" model', *Health and Social Care in the Community* 9:6 (2001), 364.

43 Mary Dixon-Woods and Richard Ashcroft, 'Regulation and the social licence for medical research', 388. This position is also in keeping with broader debates around the role of trust in regulation, with 'regulators act[ing] as third party providers of trust … in the relationship between citizens and organizations'. Frédérique Six and Koen Verhoest, 'Trust in regulatory regimes: Scoping the field', in Frédérique Six and Koen Verhoest (eds.), *Trust in Regulatory Regimes* (Cheltenham: Edward Elgar Publishing, 2017), 10.

44 Theodore Porter, *Trust in Numbers: The Pursuit of Objectivity in Science and Public Life* (Princeton, NJ: Princeton University Press, 1995). For an alternative discussion of the relationship between Porter's ideas and RECs, see Sarah Dyer, 'Applying Bioethics', 214.

45 Charles Bosk and Joel Frader, 'Institutional Ethics Committees: Sociological Oxymoron, Empirical Black Box', in Raymond DeVries and Janardan Subedi (eds.), *Bioethics and Society: Constructing the Ethical Enterprise* (Upper Saddle River, NJ: Prentice Hall, 1998), 95.

46 Office of Inspector General, Department of Health and Human Services, *Institutional Review Boards: Their Role in Reviewing Approved Research* (Washington, DC: US Government Printing Office, 1998), 12.

47 Paul McNeil, *The Ethics and Politics of Human Experimentation* (Cambridge: Cambridge University Press, 1993), 110–111.

48 For the classic discussion of varieties of isomorphism, see: Paul Dimaggio and Walter Powell, 'The iron cage revisited: Institutional isomorphism and collective rationality in organizational fields', *American Sociological Review*, 48:2 (1983), 147–160.

49 Rachel Douglas-Jones, 'A single broken thread: Integrity, trust and accountability in Asian ethics review committees', *Durham Anthropology Journal*, 18:2 (2012), 18–19.
    At first glance, the development of other oversight mechanisms – such as pharmaceutical company monitoring – might be seen as eroding the role of trust in REC decisions (since REC members might expect that any infringement of approval conditions will be detected, if not by the REC). However, in the UK the underlying culture of REC decision making developed at a time that Nigel Baber, a clinical pharmacologist with over 20 years' experience in the pharmaceutical

industry, describes as: 'the "golden days in clinical pharmacology in industry" … when we didn't require regulatory approval but we could move quickly into volunteers after local ethics committee approval. It was the raison d'être why US companies wanted clinical pharmacology to be done in their European subsidiaries … Our guardians were the ethics committees, who had to act both as ethics committees and scientific committees': in Lois A. Reynolds and E.M. (Tilli) Tansey (eds.), *Clinical Pharmacology in the UK, c. 1950–2000: Industry and Regulation* (London: Wellcome Trust, 2008), 35–36.

50 Mary Dixon-Woods, Karen Yeung, and Charles Bosk, 'Why is UK medicine no longer a self-regulating profession? The role of scandals involving "bad apple" doctors', *Social Science and Medicine*, 73:10 (2011), 1452–1459. Additional limits to an argument based purely on the importance of the medical profession can be found in Dixon-Woods and Ashcroft's point that medical research in the UK is no longer dominated by doctors, and while 'Researchers make many claims for the professional features of the work they undertake … they do not display the characteristics of a single profession.' Mary Dixon-Woods and Richard Ashcroft, 'Regulation and the social licence for medical research', 385.

51 Keith Hawkins, *Law as Last Resort: Prosecution Decision-Making in a Regulatory Agency* (Oxford: Oxford University Press, 2002), 47. Other studies support the argument that trust-based regulatory systems have certain advantages – such as better compliance rates – over more intrusive approaches: Kristina Murphy, 'The role of trust in nurturing compliance: A study of accused tax avoiders', *Law and Human Behavior*, 28:2 (2004), 187–209; Neil Gunningham and Darren Sinclair, 'Regulation and the role of trust: Reflections from the mining industry', *Journal of Law and Society*, 36 (2009), 167–194; John Braithwaite and Toni Makkai, 'Trust and compliance', *Policing and Society*, 4:1 (2010), 1–12; Michelle Pautz and Sara Rinfret, 'State environmental regulators: Perspectives of trust with their regulatory counterparts', *Journal of Public Affairs*, 16:1 (2016), 28–38.

52 Frédérique Six and Koen Verhoest, 'Trust in regulatory regimes', 10. See also: Adrian Cherney, 'Trust as a regulatory strategy: A theoretical review', *Current Issues in Criminal Justice*, 9:1 (1997), 71–84.

53 For the value of inter-organisational trust see: Jörg Sydow and Arnold Windeler, 'Knowledge, trust, and control: Managing tensions and contradictions in a regional network of service firms', *International Studies of Management & Organization*, 33: 2(2003), 69–99. For intra-organisational trust, see: Chris Grey and Christina Garsten, 'Trust, control and post-bureaucracy', *Organization Studies*, 22:2 (2001), 229–250.

54 Adam Hedgecoe, 'Trust and regulatory organizations: The role of local knowledge and facework in research ethics review', *Social Studies of Science*, 42: 5(2012), 662–683.

55 It is worth noting, however, that there is speculation that ethics review processes that are deemed to be unfair by researchers may increase the chances of unethical behaviour by researchers: Patricia Keith-Spiegel and Gerald Koocher, 'The IRB paradox: Could the protectors also encourage deceit?' *Ethics & Behavior*, 15:4 (2005), 339–349.

56 Gregort Bigley and Jone Pearce, 'Straining for shared meaning in organization science: Problems of trust and distrust', *Academy of Management Review*, 23:3 (1998), 408; see also Denise Rousseau, Sim Sitkin, Ronald Burt, and Colin Camerer, 'Not so different after all: A cross-discipline view of trust', *Academy of Management Review*, 23:3 (1998), 393–404.

57 Paul Dumouchel, 'Trust as an Action', *European Journal of Sociology/ Archives européennes de sociologie*, 46:3 (2005), 425.

58 Philip Hamburger, 'The New Censorship: Institutional Review Boards' (University of Chicago Public Law and Legal Theory Working Paper No. 95, 2005); Philip Hamburger, 'Getting permission', *Northwestern Law Review*, 101:2 (2007), 405–492; Jack Katz, 'Towards a natural history of ethical censorship', *Law and Society Review*, 41:4 (2007), 797–810.

59 Robert Dingwall, 'Turn off the oxygen ...' *Law & Society Review*, 41:4 (2007), 790; 'Confronting the anti-democrats: The unethical nature of ethical regulation in social science', *Medical Sociology Online* 1 (2006), 57.

60 Martyn Hammersley, 'Against the ethicists: On the evils of ethical regulation', *International Journal of Social Research Methodology*, 12:3(2009), 211–225; 'Creeping ethical regulation and the strangling of research', *Sociological Research Online*, 15:4 (2010), 16.

61 The earliest studies include Bernard Barber, John Lally, Julia Loughlin Makarushka, and Daniel Sullivan's *Research on Human Subjects: Problems of Social Control in Medical Experimentation* (New York: Russel Sage Foundation, 1973) and Bradford Gray, *Human Subjects in Medical Experimentation* (New York: John Wiley, 1975). Surveys of US IRB practice include: Bradford Gray, Robert Cooke, and Arnold Tannenbaum, 'Research involving human subjects', *Science*, 201: 4361 (1978), 1094–1101; James Bell, John Whiton, Sharon Connelly, and James Bell Associates, *Evaluation of NIH Implementation of Section 491 of the Public Health Service Act, Mandating a Program of Protection for Research Subjects* (Bethesda, MD: National Institutes of Health, 1998); and William

Rothstein and Linh Phuong, 'Ethical attitudes of nurse, physician, and unaffiliated members of institutional review boards', *Journal of Nursing Scholarship*, 39:1 (2007), 75–81. A good review of US-centred empirical research can be found in: Lura Abbott and Christine Grady, 'A systematic review of the empirical literature evaluating IRBs: What we know and what we still need to learn', *Journal of Empirical Research on Human Research Ethics*, 6:1 (2011), 3–19.

Surveys focusing on UK NHS RECs include: Ian Thompson, Kate French, Kath Melia, Kenneth Boyd, Alan Templeton, and Brian Potter, 'Research ethical committees in Scotland', *British Medical Journal*, 282:6265 (1981), 718–720; Pauline Allen and W.E. Waters, 'Attitudes to research ethical committees', *Journal of Medical Ethics*, 9:2 (1983), 61–65; Richard Nicholson, *Research With Children: Ethics, Law, and Practice* (Oxford: Oxford University Press, 1986), Chapter 8; Gerry Kent, 'The views of members of Local Research Ethics Committees, researchers and members of the public towards the roles and functions of LRECs', *Journal of Medical Ethics*, 23:3 (1997), 186–190; L. Elliott and David Hunter, 'The experiences of ethics committee members: Contradictions between individuals and committees', *Journal of Medical Ethics*, 34:6 (2008), 489–494; R. Hernandez, M. Cooney, Christian Dualé, et al. 'Harmonisation of ethics committees' practice in 10 European countries', *Journal of Medical Ethics*, 35:11 (2009), 696–700.

62 Robert Klitzman, *The Ethics Police? The Struggle to Make Human Research Safe* (New York: Oxford University Press, 2015), 22 and 58. Other studies that rely on interviewing members of US IRBs include: Sohini Sengupta and Bernard Lo, 'The roles and experiences of nonaffiliated and nonscientist members of institutional review boards', *Academic Medicine*, 78:2 (2003), 212–218; Emily Anderson, 'A qualitative study of non-affiliated, non-scientist institutional review board members', *Accountability in Research*, 13:2 (2006), 135–155; Scott Burris and Kathryn Moss, 'U.S. health researchers review their ethics review boards: A qualitative study', *Journal of Empirical Research on Human Research Ethics*, 1:2 (2006), 39–58.

63 Damon Parker, Michael James, and Robert Barrett, 'The practical logic of reasonableness: An ethnographic reconnaissance of a research ethics committee', *Monash Bioethics Review*, 24:4 (2005), 8.

64 Howard Becker and Blanche Geer, 'Participant observation and interviewing: A comparison', *Human Organization*, 16:3 (1957), 30–31.

65 Colin Jerolmack and Shamus Khan, 'Talk is cheap: Ethnography and the attitudinal fallacy', *Sociological Methods & Research*, 43 (2014), 179.

66 Robert Klitzman, *The Ethics Police?*, 22. Of course, it is not that observational data provide a 'transparent' account of REC decision making. Fieldnotes are partial accounts, reflecting the interests and views of the ethnographer.

67 For example, Klitzman's work draws mainly on 46 IRB interviews, where 28 interviewees (61%) were chairs/co-chairs, ten (22%) administrators, and only seven ordinary members (15%) (23). Surveys can be even more exclusive, with, for example, Amy Whittle and colleagues' work drawing on information from a telephone interview survey of 188 IRB chairs: Amy Whittle, Seema Shah, Benjamin Wilfond, Gary Gensler, and David Wendler, 'Institutional review board practices regarding assent in pediatric research', *Pediatrics,* 113:6 (2004), 1747–1752. UK-based surveys tend to poll all members within specific geographical areas or of particular committees – perhaps a consequence of a more centralised system making access to non-chair members easier.

68 Richard Nicholson, 'What do they get up to? LREC annual reports', *Bulletin of Medical Ethics,* 129 (1997), 13–23; Emma Godfrey, Emma Wray, and Richard Nicholson, 'Another look at LREC annual reports', *Bulletin of Medical Ethics,* 171 (2001), 13–21.

69 Emma Angell and Mary Dixon-Woods, 'Do research ethics committees identify process errors in applications for ethical approval?' *Journal of Medical Ethics,* 35 (2009), 130–132; Emma Angell, Alan Bryman, Richard Ashcroft, et al., 'An analysis of decision letters by research ethics committees: The ethics/scientific quality boundary examined', *BMJ Quality & Safety,* 17 (2008), 131–136; Mary Dixon-Woods, Emma Angell, Richard E. Ashcroft, and Alan Bryman, 'Written work: The social functions of Research Ethics Committee letters', *Social Science & Medicine,* 65:4 (2007), 792–802; Justin Clapp, Katherine Gleason, and Steven Joffe, 'Justification and authority in institutional review board decision letters', *Social Science & Medicine,* 194 (2017), 32.

70 Sheila Jasanoff, *The Fifth Branch: Science Advisers as Policymakers* (Cambridge, MA: Harvard University Press, 1990); Roger Smith and Brian Wynne (eds.), *Expert Evidence: Interpreting Science in the Law* (London: Routledge, 1989); A. Irwin, H. Rothstein, H.S. Yearley, and E. McCarthy, 'Regulatory science – towards a sociological framework', *Futures,* 29:1 (1997), 17–31; Alberto Cambrosio, Peter Keating, Thomas Schlich, and George Weisz, 'Regulatory objectivity and the generation and management of evidence in medicine, *Social Science & Medicine,* 63:1 (2006), 189–199.

71 J. Murphy, L. Levidow, and S. Carr, 'Regulatory standards for environmental risks: Understanding the US–European union conflict over

genetically modified crops', *Social Studies of Science,* 36:1 (2006), 133–160; John Abraham and Courtney Davis, 'Deficits, expectations and paradigms in British and American drug safety assessments – prising open the black box of regulatory science', *Science Technology & Human Values,* 32:4 (2007), 399–431; Joan Robinson, 'Bringing the pregnancy test home from the hospital', *Social Studies of Science,* 46:5 (2016), 649–674; Margaret Sleeboom-Faulkner, 'Regulatory brokerage: Competitive advantage and regulation in the field of regenerative medicine', *Social Studies of Science,* 49:3 (2019), 355–380.

72 Recent examples include: A.L. Roth, J. Dunsby, and L.A. Bero, 'Framing processes in public commentary on US federal tobacco control regulation', *Social Studies of Science,* 33:1 (2003), 7–44; Alina Geampana, 'Risky technologies: Systemic uncertainty in contraceptive risk assessment and management', *Science Technology & Human Values,* 44:6 (2019), 1116–1138; Shai Mulinari and Courtney Davis, 'The will of Congress? Permissive regulation and the strategic use of labeling for the anti-influenza drug Relenza', *Social Studies of Science,* 50:1 (2020), 145–169.

73 For an exception to this approach, see: Patrick Brown, Ferhana Hashem, and Michael Calnan, 'Trust, regulatory processes and NICE decision-making: Appraising cost-effectiveness models through appraising people and systems', *Social Studies of Science,* 46:1 (2016), 87–111.

74 Laura Stark, *Behind Closed Doors: IRBs and the Making of Ethical Research* (Chicago, IL: University of Chicago Press, 2011); 'IRBs and the Problem of Local Precedents', in I. Glenn Cohen and Holly Fernandez Lynch (eds.), *The Future of Human-Subjects Protections* (Cambridge, MA: MIT Press, 2014); 'IRB Meetings by the Minute(s), How Documents Create Decisions', Charles Camic, Neil Gross, and Michèle Lamont (eds.), *Knowledge Making, Use and Evaluation in the Social Sciences: The Underground of Practice* (Chicago, IL: The University of Chicago Press. 2011).

75 Julie Morton, ' "Text-work" in research ethics review: The significance of documents in and beyond committee meetings', *Accountability in Research,* 25:7–8 (2018), 387–403; Jean Philippe De Jong, Myra Van Zwieten, and Dick Willems, 'Ethical review from the inside: Repertoires of evaluation in research ethics committee meetings', *Sociology of Health & Illness,* 34:7 (2012), 1039–1052 and Patricia Jaspers, Rob Houtepen, and Klasien Horstman, 'Ethical review: Standardizing procedures and local shaping of ethical review practices', *Social Science & Medicine,* 98 (2013), 311–318; Damon Parker, Michael James, and Robert Barrett, 'The practical logic of reasonableness'.

76 Edward Dove, *Regulatory Stewardship of Health Research.*

77 Adam Hedgecoe, 'Trust and regulatory organizations: The role of local knowledge and facework in research ethics review', *Social Studies of Science*, 42:5 (2012), 662–683; Laura Stark, 'Reading trust between the lines: "Housekeeping work" and inequality in human-subjects review', *Cambridge Quarterly of Healthcare Ethics*, 22:4 (2013), 391–399; Edward Dove, *Regulatory Stewardship of Health Research*.

78 The RECs have been given pseudonyms, and some of their details have been altered to preserve committee and members' anonymity. This extends to the ethnographic fieldnotes presented below: details of specific treatments and research have been deliberately obscured.

79 In December 2006 (i.e. just after my fieldwork), there were 14 so-called Type 1 RECs (not represented in my observational sample); 55 Type 2 committees (like St Swithin's LREC); and 47 Type 3 RECs (like Northmoor LREC or Coastal MREC). The committees included in my research therefore carry out the same role as the majority of RECs in England at the time. This typology of RECs has steadily eroded over time, with the only distinction now drawn between Type 1 committees which can review Phase I clinical trials (as well as other applications) and Type 3 committees which can review all other kinds of application apart from Phase I trials.

80 Maureen Fitzgerald, Paul Phillips, and Elisa Yule, 'The research ethics review process and ethics review narratives', *Ethics & Behavior*, 16:4 (2006), 377–395; Maureen Fitzgerald, 'Punctuated equilibrium, moral panics and the ethics review process', *Journal of Academic Ethics*, 2:4 (2005), 315–338; Sarah Dyer, 'Applying Bioethics'.

81 For example, when discussing the attendance of 'a newly appointed … local manager' at a meeting, Sarah Dyer notes that: 'The committee seemed unsure whether they were being checked up on or not and were much more wary and tentative in their discussions *than other committees I had seen*' (emphasis added). With only one observation at this REC, Dyer cannot draw a comparison to how this committee normally behaves (i.e. without the presence of the manager), and is forced into rather speculative comparison with other RECs: Dyer, 'Applying Bioethics', 80.

82 Fitzgerald, 'Punctuated Equilibrium'.

83 Michael Bacharach and Diego Gambetta, 'Trust in Signs', in Karen Cook (ed.), *Trust in Society: Volume II in the Russell Sage Foundation Series in Trust* (New York, NY: Russell Sage Foundation, 2001), 150.

84 Diego Gambetta and Heather Hamill, *Streetwise: How Taxi Drivers Establish Customers' Trustworthiness* (New York, NY: Russell Sage Foundation, 2005), 6.

85 Michael Bacharach and Diego Gambetta, 'Trust in Signs', 154.

86 For criticism of cognitive models, see Oliver Williamson, 'Calculativeness, trust and economic organization', *Journal of Law and Economics*, 26 (1991), 453–486; but see also David Lewis and Andrew Weigert, 'Trust as a social reality', *Social Forces*, 63:4 (1985), 967–985.

87 For examples of press coverage see: Polly Curtis, 'Drug trial company provisionally cleared by medicines regulator: Disaster could not have been foreseen, says report', *The Guardian* (April 6 2006), 8; David Rose, 'Anger as expert report into drug trial disaster fails to allot blame', *The Times* (December 8 2006), 34. For analysis of press coverage, see Lynne Stobbart, Madeleine Murtagh, T. Rapley, Gary Ford, et al., 'Medicine and the media: "We saw human guinea pigs explode"', *British Medical Journal*, 334 (2007), 566–567. A medical explanation for what happened can be found in: Ganesh Suntharalingam, Meghan Perry, Stephen Ward, Stephen Brett, Andrew Castello-Cortes, Michael Brunner, and Nicki Panoskaltsis, 'Cytokine storm in a Phase I trial of the anti-CD28 monoclonal antibody TGN1412', *The New England Journal of Medicine*, 355:10 (2006), 1018–1028; Richard Stebbings, Lucy Findlay, Cherry Edwards, David Eastwood, et al., '"Cytokine Storm" in the Phase I trial of monoclonal antibody TGN1412: Better understanding the causes to improve pre-clinical testing of immunotherapeutics,' *Journal of Immunology*, 179 (2007), 3325–3331.

88 For an analysis of the role of REC review and the regulatory culture in approving short internal dosing: Adam Hedgecoe, 'A deviation from standard design? Clinical trials, research ethics committees and the regulatory co-construction of organizational deviance', *Social Studies of Science*, 44:1 (2014), 59–81.

89 While this committee was formally called (for unclear reasons) the Brent Medical Ethics Committee (MEC), for clarity this book refers to it as the Brent Research Ethics Committee (REC).

90 Diane Vaughan, *The Challenger Launch Decision: Risky Technology, Culture and Deviance at NASA* (Chicago, IL: University of Chicago Press, 1996); 'The dark side of organizations: Mistake, misconduct, and disaster', *Annual Review of Sociology*, 25 (1999), 271–305.

1

Paper promises or written applications as trust warrants

'Someone unable to put together a competent ethics submission is almost certainly unable to do competent ethical research.'
Dr Tim Steiner, Chair, London Multi-centre
Research Ethics Committee (MREC)[1]

'Sloppy proposals, in Dr. M's view, were written by sloppy investigators, who had the potential to threaten the well-being of research participants.'
Laura Stark[2]

Amy leads off on an application to Northmoor Local Research Ethics Committee (LREC). The proposed research involves the development of a survey of mental health inpatients. Her main concern centres on the six attempts to contact people to ask them to fill in the questionnaire which is described as 'hectoring': 'There is an issue around recruiting people with mental health problems but six attempts is hounding them.' Overall, her position is that 'In principle it's a good idea but they need to do a bit of work. In the PIS [participant information sheet] they don't really say what they're going to do and why.'

The discussion opens up to the rest of the committee. Rose, the REC's statistician, suggests that for her the issue 'is whether the results are reliable. I also think a 30% response rate is alright. They could also analyse non response in terms of institutions, but I'm not sure the results will mean anything.' Amy chips in: 'The other thing is they don't report the results back to the patients and given they contact them six times to get them to fill out the form, the least they could do is send a report.' Points begin to emerge as members take turns to criticise the application:

Damien: 'I can't believe that they got quarter of a million pounds to do this!'

Amy: 'And the Hospital Trust is giving the names and
            addresses to the researchers, rather than sending
            out the questionnaire itself: I think I've talked
            myself out of it! I was feeling a lot more charitable
            this morning ...'
Damien: 'Question 3 [in the questionnaire] looks like it's
            asking 'did you sleep with any other patient?'
Charlotte: 'Is it doing anything except bothering people six
            times?'
Rose: 'Being contacted six times could be harmful to
            people with mental health issues.'
Amy: 'It's a wicked waste of money.'
Frank: 'It may well be but we can't refuse it on those
            grounds.'

Because the applicant has not come along to the meeting to answer questions, the REC moves on to discussing its final decision.
Amy sums up her position (more sceptical than at the start of the
discussion): 'I think we should say it's unsuitable, so they come
back and have a talk about it.' Referring to the concerns expressed
about this research as a waste of money, Frank, the Vice Chair who
is running this meeting, reiterates the point: 'we can't reject it on
those grounds – we need a reason'.

Amy: 'Oh, I've got reasons!'
Amanda: 'The main one being the re-contact.'
Damien: 'If they are trying to look at improving the response
            rate they don't need this number of people.'

Yet for some members, concerns about repeated re-contact are
not enough to fully reject an application (as opposed to just requiring some changes):

Neil: 'But we can't reject it on these grounds.'
Frank: 'I agree.'

In response, Rose restructures the objection to the proposal,
moving beyond issues about re-contact or it being a waste of
funding to deeper concerns about the underlying justification: 'But
they haven't given a proper reason for doing it. Improving response rate is not a good enough reason.' Like the large majority

of applications to RECs, this study is finally given a 'provisional' approval, requiring additional information or some minor changes, before being signed off by the Chair or a small sub-group of the REC.

As this example suggests, the application form and associated pieces of paper are the cornerstone of any REC decision; they provide the raw material for any assessment of the suitability of a piece of research and the trustworthiness (or otherwise) of an applicant. This chapter sets out to explore the way in which RECs draw on written materials submitted by applicants to make initial (sometimes final) decisions about whether to approve a piece of proposed research or not, and, in the process, make judgements about whether an applicant can be trusted to do what they say they will do. A key insight is into the tangled relationship between the idea of written consent to research and the rise of REC review and how consent documents (PIS, informed consent forms) serve as a form of limit to REC trust decisions.

At the most basic level the application form is obviously a source of information about the proposed study, explaining what will be done, to whom, for how long and with what possible risks and potential benefits. As such, this dry, standardised document might be thought of as largely devoid of 'social' information. Yet for members of RECs, the written application is also an important source of intelligence on the researcher him- or herself. As Adam, a lay member of the Coastal MREC, put it: 'If somebody's put in a pretty awful application, it's quite clear they're not really interested in what the participants might be thinking or feeling ... the fact that they haven't done the form properly, in a sense is indicative of a cavalier attitude.'

At the same time, a well-completed application form, and associated material, will reassure a REC of an applicant's trustworthiness. For example, during an assessment of a study of the care of dementia patients, Coastal MREC discusses the methods being proposed, with Adam noting that 'given the detail they have given in terms of instructions for those people filling out the questionnaires, I don't think we need to be concerned about the focus groups'. While the nature of the personal relationship with applicants is an important part of REC decision making (see Chapters 3 and 4), the use of application forms to pass judgements underlines how documents allow a 'virtual relationship' to development between RECs and applicants. Yet the interpretation of the application and the risk of the intervention sit within a broader social context, including expectations of trustworthiness, which shape the reading of the application, regardless of personal interaction with applicants.

That RECs are bodies that produce, manipulate, and distribute large amounts of paper is acknowledged both by the committees themselves – Martin, Chair of Northmoor REC, sarcastically calling across the pile of paper in front of him at the beginning of a meeting: 'welcome to the Amazon rainforest preservation society' – and by applicants who regularly complain about the amount of paper that is required for an application to a National Health Service (NHS) REC.[3] This latter point highlights the paradox that, as is a common feature for changes in the REC system, the original drive for the standardisation of the NHS form was the desire to make things easier for researchers. In the mid-1990s the head of one of the NHS Research and Development (R&D) Directorates – units with responsibility for oversight and support of research at a regional level – was a public health epidemiologist who, because of his own research experience, was aware of the difficulty of setting up large studies across different ethics committees and getting contradictory opinions, as well as the challenge of filling in several different application forms. As a consequence, meetings with RECs in that part of the country were organised and their various application forms pulled together. And, as one person involved in this process put it, the result was:

> something that most of them were happy to adopt ... we had it nicely designed, it looked nice, it had a proper layout, it was sensible, it was comprehensive. So, that was a great achievement, actually. We had a single application form across the whole of the [region] that all the ethics committees were using and you just had to put a different ethics committee name on it ... And I'd like to think it improved the process of ethical review because it was structured and standardised. You know, it had specific sections about interventions and risks and benefits and stuff.

However, when this regional form was adopted at a national level it became, over time, more cumbersome as additional sections and remits were added to it: 'It all got more complicated then because of the research governance agenda, the [Human] Tissue Bill and data protection and genetics. So, it, it became rather mammoth which is sort of gradually being reformed down again, I think, but it was quite difficult to contain it.'

From a Science and Technology Studies (STS) perspective, thinking about application forms and their role in trust decisions most obviously draws on Steven Shapin's concept of a 'literary technology of trust'. In his classic discussion of Robert Boyle's attempts to persuade his seventeenth-century contemporaries of the value of his experiments with an air pump, Shapin argues that Boyle sought to draw up the rules for scientific reports, creating a 'literary technology: the expository means by which matters of fact were established and assent mobilized'. The aim of such written reports was to give the reader the impression that they had actually seen Boyle's air pump work – the experience of 'virtual witnessing' – essentially creating 'a technology of trust and assurance that the things had been done and done in the way claimed'.[4]

Of course, for some commentators, basing trust decisions on what people write down is unsatisfactory. As the leading US bioethicist Arthur Caplan noted in the early 1980s:

> The IRB process ... ultimately depends on the good-faith compliance of those researchers it is charged with supervising ... [As a process] ... it devotes too much time to the production of paper promises and almost no time to the enforcement, investigation or general assurance that the promises will be kept.[5]

Yet Caplan's position – unsurprising for a bioethicist, perhaps – lacks awareness of the social information contained in such documents. While the specific trust-generating elements of Boyle's literary technology (prolixity, iconography, and modesty) may not be relevant to REC application forms, the fundamental point – that readers of even the most dry and formal written documents make trust decisions about authors – is a key insight, and one common in broader discussion of regulatory decision making. As Heimer and Gazley note in their work on the regulation of HIV clinics, people's ability to complete application forms caused regulators to wonder 'If clinic staff cannot manage such mundane chores as making copies in triplicate ... how can they possibly manage the more complicated tasks of good science and good caregiving?'[6]

This chapter makes the case that in the context of research ethics applications, written materials allow the 'virtual witnessing' of both applicants' character (whether they are arrogant or not, for instance) and, moving beyond just Shapin's argument about convincing accounts of past behaviour (this is how the experiment was done), their future behaviour (how carefully they will seek consent from patients, for instance). This is then used as the foundation for a trust decision about the applicant, to which other forms of trust warrant – from previous 'local knowledge' of an applicant (Chapter 2), or how they present themselves in front of the committee when invited to answer questions (Chapter 3) – are added. In terms of idiocultural variation between committees, while ethics committees' assessment of these application forms do vary, the format of NHS REC meetings – guided as they are by a set of standard operating procedures (SOPs) – tend to resemble each other.[7] While, as a result of the legislation around pharmaceutical trials, the documents and rules surrounding REC review tend to divide between those focused on Clinical Trials of Investigational Medicinal Products (CTIMPs – 'drugs trials') and those focused on all other research, the way in which REC meetings proceed is largely blind to this distinction. Members arrive at the meeting room having been sent the relevant paperwork about a fortnight before, having already formed a first impression about the application. However the REC's discussion goes and whether or not the applicant attends, the form and other written documents serve as an anchor for individual members' decisions about an application.[8] What people actually receive varies according to their role and expertise. All members will get application forms and consent documents (the PIS and the consent form), but only the pharmacist on the committee and the chair are likely to be sent the Investigator's Brochure (IB: a comprehensive summary of information about a drug) for a drug trial.

Following introductions and any relevant news (proposed policy changes around REC review perhaps), a REC meeting starts by reviewing applications. The committees I observed used a 'lead reviewer' system where two members of the committee were flagged as having responsibility to lead discussion on an application. While other members are expected to have read the application form and materials, the lead reviewers provide a brief synopsis of the proposal and an outline of the perceived ethical issues. Discussion then broadens out to the wider committee. The choice of who is picked as leads is usually made by the Chair and administrator, on the basis of expertise and type of member. The two LRECS I observed tended tend to choose an expert and a lay member to share the lead duties on most applications, with the expert member being selected due to 'fit' with the research topic – for example, an oncology nurse will be the lead on a cancer drug trial, a psychiatrist on research on new treatments for schizophrenia. Coastal MREC, in contrast, tended to use only a single lead reviewer, with some fit to expertise, but sharing the role between experts and lay members.

As discussion of an application moves round the table, the REC steadily accumulates a list of questions, concerns, or points it would like to see clarified. The skill of the Chair is to know when to draw these discussions to a halt and move on to the next stage, and when to allow the talk to continue. As Colin, the lay Vice Chair of Coastal, notes:

> what a chair does need to be able to do … is to spot the moment when you can offer the possibility of decision and it seems to me there are critical points the discussion reaches and that's a chairing skill when you might want to go, as it were, for a 'window' and say are we ready for a decision now?

Eventually, the Chair will normally suggest that the applicants, who will have been invited to the meeting by letter, are asked to come into the room to respond to the REC's concerns. If the applicants are sitting outside they will come in and engage with the REC (see Chapter 4). If not the committee will move on to making a decision about the application, whether to approve it, reject it or, as in the majority of cases, give a provisional approval (i.e. to approve it provided specified changes are made – termed *favourable opinion following provisional*).[9] Practice then varies between committees: St Swithin's and Northmoor LRECs move on to the next application, while in Coastal MREC meetings the administrator pops out to speak to the applicants, who will have been asked to wait outside, to let them know the outcome.

For REC members at least, it is clear that the face-to-face, collegial nature of REC meetings is an important part of the review process, often claiming that the meeting brings out the complexities of applications that they, as individual reviewers, felt were straightforward. Abigail, from St Swithin's, notes that she is 'always surprised by the way in which projects that I read through and thought nothing about, somebody raises the point that the committee then will discuss for ages and I thought it would go through 'on the nod' [i.e. with little debate]'. At one level this is about the support of having other people review an application at the same time as yourself. In comparison to reviewing in the sub-committee context where individual members sign off on proposed amendments to already approved applications – 'it's horrible doing it on your own because there's a sense you'll get it wrong and there's somebody's research resting on it and you don't know all of it' – Louise (St Swithin's LREC) suggests that 'what's nice about doing it at the Committee, is if you've got it wrong, somebody else would have spotted it'.

Yet in addition to such 'safety net' arguments, deeper benefits are claimed for physical meetings, with Malcolm, the Chair of Coastal, suggesting:

> not infrequently an application comes up which you personally feel is going to be approved in a straightforward manner. While sometimes people disagree and raise points you think are unnecessary, as often as not somebody says something and you might think, 'oh well, that's a good point. I hadn't thought about that' and that's the beauty and the strength of the committee.[10]

Given the perceived importance of face-to-face REC meetings, it is perhaps surprising that it has only been a standard part of REC practice for a relatively short period of time. For example, of the 18 RECs represented at a meeting organised by the West Birmingham Community Health Council, in 1988 (i.e. 20 years after RECs had been rolled out across the country), one finds a wide variety of practice with regard to physical meetings, with some committees meeting once a month or even more (e.g. Nottingham City REC at around 20 meetings a year) and others meeting every couple of months or even not at all.[11] As we might expect, as standardisation of REC practice began to develop in the early 1990s, *not* holding REC meetings became seen as increasingly unacceptable. Arguing that 'A disadvantage of working by post is that Committee meetings may become so rare that the valuable mutual exchanges between members are lost, and lay members particularly will feel isolated', the disapproval in the Royal College of Physicians' (RCP's) first set of guidelines from 1984 – 'it is inappropriate to seek to conduct all business without meetings' – escalates in their revised guidelines six years later: 'Reasonably frequent meetings are essential to allow a Committee ethos to develop. To work entirely or almost entirely by mail or by chairman's decision ... is unacceptable ... it is inappropriate to seek to conduct all business without meetings.' The Department of Health's 'Red Book' in 1991 is more moderate – 'Conducting business by post or telephone should be discouraged' – although the standards and training documents produced by the Department of Health at around the same time emphasise the central role of face-to-face meetings.[12]

In terms of assessing the written application – and, implicitly, the trustworthiness of the applicant – the first point of entry for much discussion on the part of a REC is the clarity of the information on the form – can the committee members understand what is going to be done and why? For REC members, lack of clarity is a broad category, covering a wide range of errors such as: technical jargon, intra- and inter-document inconsistency; or a lack of detail regarding the medical interventions that will be used. A good example of the way in which lack of clarity produces concern on the part of REC members can be seen in the case of a proposal for work on adult stem cells reviewed by St Swithin's LREC. The application is introduced by Roderick, who describes himself as 'not happy about this at all' since the applicant doesn't say what he's going to do with the results of his investigations. The applicant says he will use 'historical' samples, perhaps from a tissue bank (although this is not clearly spelled out) but he also wants to get tissue from current and future patients, yet there's no mention of informed consent, underlining the importance of a clear application when it comes to building trust about future behaviour. The application claims that the proposal has had both internal and external review yet the applicant attaches no referee's comments (although the form requests this). The study is vague on statistics. 'The whole thing is quite laughable', and 'I'm surprised [the department head] put his name to it.' In an unusually strong statement, Roderick concludes by saying that this study is a 'totally inadequate application and needs to be rejected!'

Dean points out that the applicant 'may be here!' (i.e. outside the door waiting to attend), and Abigail mischievously turns to Roderick and says: 'This is where you tell him!' (i.e. to his face). Roderick smiles and looks a bit sheepish. Dean then offers his own comments, muttering something about the applicant apparently trying to seek consent from patients going under anaesthetic, again highlighting concerns about future behaviour. Looking up from the application, Ingrid then says: 'his background ... he has no biology background at all. If he's never worked in a lab, is it ethical to do patient samples?' Abigail replies: 'I don't think he'll like a question like that', to which Ingrid replies: 'well it needs to be phrased more tactfully than that ...' before then raising the scientific issue of whether the tissue the applicant is interested in (tumour tissue) would actually contain the cells he wanted to sample – 'The science is non-existent.' While the REC waited to see if the applicant was going to turn up in person, it turns to another study. The applicant does not, in the end, turn up and the application is rejected ('unfavourable opinion'), with a set of suggested changes compiled to send to the researcher.

As we will see in Chapter 3, the presentation of information clearly and in comprehensible terms is important when applicants appear before a REC to answer questions, but such clarity is also crucial in the written application.[13] Failure to explain what one is planning to do is easily interpreted by REC members in terms of an applicant's character, as happened with another application to St Swithin's involving use of data from the National Blood Service (NBS) where Ingrid interpreted the application as being presented in an arrogant manner: 'I assumed they were just being dismissive and not wanting to answer our questions.' Things are complicated because of the nature of the study; it is not clear that it actually counts as research (as opposed to REC-exempt clinical audit) and the committee may not need to see the application.[14] Yet the lack of clarity means that, according to Ingrid, 'As it stands it's un-assessable.' Ben, the committee's administrator, says that the Central Office for Research Ethics Committees (COREC: the Department of Health's unit that coordinates RECs across the NHS), having been consulted, require the study to come to a REC. Ingrid reiterates the point that the application is 'a touch on the arrogant side' and Dean, the committee's lay Vice Chair who is standing in as Chair for this meeting, summarises by saying, 'Send it back, rewrite it in a less arrogant manner', making clear how the study conforms to the NBS's ethical standards.

There are other ways in which the assessment of this application could have gone. The uncertainties around whether it needed
to come to REC review at all could be interpreted as playing a more
important role in shaping the applicant's lack of clarity or could lead
to a more generous interpretation of an application that is seen, instead, in terms of arrogance and unwillingness to engage with the
REC properly ('just being dismissive and not wanting to answer our
questions'). As Yvonne, a lay member of Coastal MREC, put it, for
committee members, 'in an awful lot of research' lack of clarity is
interpreted as 'just laziness, as far as some researchers are concerned,
or thoughtlessness. They just haven't thought it through … so I sort
of, get irritated they haven't thought it through.'[15] Yet this 'reading of
character' – and hence trustworthiness – into written applications is
not just a feature of anticipatory ethics review but is a characteristic
of regulatory oversight more generally. For example, in the case of
the inspection of HIV research clinics Heimer and Gazley note that:

> Because it is difficult to assess compliance with deeper ethical and
> moral obligations, regulatory inspectors often treat technical compli-
> ance as an indicator of deep compliance. If staff keep their research re-
> cords properly, have supporting source documents and signed consent
> forms for all research subjects, can document that they have reported
> all serious adverse events, and so forth, then this technical compliance
> and their apparent seriousness of purpose are taken as signals that
> they almost certainly also are compliant at a deeper level.[16]

Such decisions are always made in the context of (variable) uncertainty, and as a result RECs draw on a range of broader social expectations to 'fill in the gaps' in applications. For example, in the case
of the TGN1412 application, it is clear that, challenged by uncertainties about the riskiness of the drug being tested, the Brent REC drew
on a set of pre-existing expectations about the nature of the companies that develop such products. While the REC did not necessarily
regard the TGN1412 trial as risky, there was obvious uncertainty on
the part of the committee; as one interviewee from Brent REC put it,
although 'not an immunologist but what I do know is that immunological reactions and cascades are all quite complicated' was 'very
confident reviewing … [even] … small molecules … I've got physiology and all of that I can understand'. In the case of this application,
for this member, the immunological information seemed:

all very reassuring – nothing bad happened to any of the animals and they had used … various animals including one which had the same CD, whichever it was, I can't remember … as was in humans, so it was the most similar but it wasn't a human.

However, despite this, this interviewee admitted that:

what I wasn't confident about was how much diversity is there in immune responses. So I immediately felt quite out of my depth and normally having sat on the committee, I have been on the committee for two years … Normally I'm one of the more expert people on the panel because I've got both a medical background and a research background, so I feel quite confident explaining to the other people things that I'm not happy or happy with there, and I just didn't feel that at all. So when I turned up at the meeting one of my first comments was that I said, 'I feel a bit uneasy about this because it all looks very good but I'm not sure because I'm not an immunologist, I don't know', and then I asked what [member name], who is in rheumatology … thought and he made some comments because as a rheumatologist he does a lot more with immunology.

As Chapter 3 sets out, a key aspect in persuading the committee of the safety of the trial was the confident 'performance' offered by the representative of TeGenero, the drug's developer. Yet the interpretation of the application and the risk of the intervention sit within a broader social context, including expectations of trustworthiness, which shape the reading of the application, regardless of personal interaction with applicants. In the context of TGN1412 – a drug developed by a small, inexperienced start-up company – this is most obvious in the trust assumptions made about the kinds of companies that develop drugs, normally, as one member of Brent REC put it, 'it's big drug companies and you kind of assume that they've been doing this for 75 years, that they've got pretty good checks and measures and they've got shareholders they're answerable to, to make sure things don't go wrong'. For REC members, this kind of confidence centres on these companies'

institutional experience. You kind of know that Pfizer have been doing phase one trials for decades and that the people running them have been doing them for decades and have learned a lot. And that, you know, it's not something you consciously think about it but that's actually part of the safety, is that these people have built careers doing this.

For this member of Brent REC, the obvious contrast is with TeGenero which, after the events at Northwick Park, turned out to be 'a small start-up company, they've never done a phase one trial, there were no clinicians within their group'. The point is not that TeGenero misled the REC but rather that the gaps in the information required in the application form – for example, about the background and clinical competence of the company developing the drug – required filling in, and to do this the REC fell back on assumptions drawn from its own experience and broader knowledge.

In terms of information that is included in specific sections of the application form, a key trust warrant is how the applicants respond to the question on the REC application form used at the time of my fieldwork numbered 'A68': 'What do you consider to be the main ethical issues which may arise with the proposed study and what steps will be taken to address them?' Despite being relatively innocuous-looking, this question is an important marker for committee members, seen as telling them something about the applicant – in essence, how seriously they take the process of applying for REC approval. Leaving this section of the form blank, or simply suggesting that 'There are no ethical issues in this study' (as some applications do) alerts members to an applicant's casual attitude towards the ethics approval process. Commenting on one application, Daniel (Coastal MREC, expert), noting that the applicants had filled in A68 with 'ethical issues: nil' responds: 'That began to worry me. If they don't think there are ethical issues where people receive lung biopsies then I don't know where there are.' In another case, at St Swithin's LREC, applicants apparently fail to understand the ethical issues associated with their proposal since they only list 'data protection' as a response to A68. In a Coastal MREC discussion of a Phase I study in solid tumours Leonard suggests that the proposal 'seems a well thought out phase I study but raises the normal issues of a phase I trial in terminal patients. A68 says no ethical concerns which is rather worrying!'

Such a marker is unlikely, on its own, to lead to the rejection of an application, but rather will tilt the balance of a committee's approach away from its default setting of support and trust, and towards a more sceptical position. As Daniel, an expert member of Coastal MREC, put it: 'it's one of the ones I always think about … if the A68 says nothing, I assume I now have to be persuaded it should be approved. 'Cos if anyone thinks there are no ethical issues ...'. Filling out A68 with a list of possible ethical issues, however obscure and unlikely, displays agreement with the values that underpin research ethics review; i.e. that research can harm participants, and that researchers should reduce and/or articulate those possible harms. Occasionally, when an applicant is over-zealous in filling in this section of the form, listing extremely unlikely events as possible issues, committee members might mutter about applicants 'throwing in everything but the kitchen sink', but are unlikely to read this as a sign that the applicants should not be trusted – it would not contribute to an application being rejected.

In broader terms, a second aspect of the form that RECs pay attention to is how the application conforms to some UK-specific 'stylistic' expectations. A good example of this is an applicant's willingness to translate the PIS and consent form into other languages, on the assumption that the populations being recruited from could contain people for whom English is not their first language. When discussing an application before St Swithin's LREC, Abigail quotes the form: 'it says communication problems' as one of the exclusion criteria, asking whether this 'this is a sneaky way of excluding non-English speakers?', reiterating the REC's previous discussions on the rights for non-English speakers to contribute to their community through taking part in research. On another occasion she questioned the comment on translators – ' "where translators are available, the PIS will be translated" ... What happens if a translator is not available?' Linda, another member of St Swithin's, suggests particular applicants are 'a bit arrogant about why they are only using English speakers – they don't want them [i.e. non-English speakers] because it's a common condition and they get enough cases anyway'.

From a researcher's perspective, the requirement to translate consent materials into other languages is a frustrating burden – a considerable financial challenge. As one interviewee, well acquainted with the clinical trials industry, informed me:

> So those translations cost about £75, that's cheap, per A4 page. If you've got an eight page patient information sheet that's a lot of money. And whilst they [i.e. REC] think that pharma companies are more or less like a gravy train, there are also other organisations like the MRC [Medical Research Council] that are having

to accommodate ethics committees on this and some of them and Cancer Research UK and people like that and it's just ... so basically that we can't afford to do it in the UK. If you go to France and ask them what's, what translations they do, they laugh at you. 'French' that's the answer.

And comparisons with France, as a country where such translation would not be needed, cropped up in interviews with REC members, who acknowledged the problems with insisting on translation of participant materials. For example, Craig (St Swithin's), while discussing the need to integrate non-English-speaking people into the research population, noted that:

> I think it's very complicated. I don't think you can just say, 'sod it, everyone will speak English'. I mean, the French have always taken that view. If you live in France, you speak French otherwise bugger off. We've never really taken that view and there is a non-uniform take-up of integration.

As Adam (Coastal MREC) put it: 'My guess would be that in France, the notion of translation doesn't come in, because the French ... Because every citizen in France is expected to speak French.'[17]

Sometimes the reasons given for requiring translation are local; the reception area of St Swithin's hospital lists the languages the translation service can work in for clinical services, which run into the 20s, and, as Roderick notes, the issue of translating *research* material

> applies very much to [St Swithin's] where there's very, very mixed population and you've got schools with 38 languages being spoken. I'm sort of quite used to that, living in an area like that ... it's a very difficult subject and I think applicants have to think about it because it does actually apply and to do something about it if it does actually apply. I think that the other sections of the community should not be disadvantaged because they can't take part in that but it's even more difficult for them.

Yet other REC members make the case for translation on broader, less localised grounds. Yvonne (Coastal MREC) maps out the issues clearly, ranging from the need for researchers to incorporate funding for translation into their grant applications lest it be thought that 'They just haven't thought it through', through the way in which limiting the population recruited from 'undermines the scientific validity of the topic, because it could mean that certain groups would therefore be excluded, whose very inclusion might very well come up with different results' to the more generic point that:

> everyone should have the right to be able to take part in research, because there's all sorts of pluses you can get from being in research, partly from … you can be looked after better, as a research participant, or you can just get that warm cuddly glow from, you know, I'm helping to do this, so why should some people be ruled out of getting the warm cuddly glow?

And it is this latter point, about equality of opportunity to take part in research, that REC members conceive in specifically British terms – 'I suppose it's a very British thing, isn't it?' (Caroline, Coastal MREC) – as a deep-seated 'sort of part of the culture, isn't it?' (Donald, Northmoor & District LREC). As Colin (Coastal MREC) puts it, 'I think it's partly the tradition of liberalism and respecting the other guy and recognising you have some responsibility "for strangers in your midst". You have some responsibility for that.' Thus while RECs are willing to make exceptions to the translation requirement – 'We tend to relax it, I think, where it's a smaller piece of research, and where we feel that, actually, people wouldn't have the wherewithal to do it' (Rachael, Coastal MREC) – failing to offer to translate participant materials can be seen as failure to engage with deep-seated British cultural values, and hence to raise questions about a researcher's trustworthiness. Given the way in which REC members are drawn from a very specific section of the UK population (middle-class professionals), we might expect them to thus have a very specific – liberal, multicultural perhaps – interpretation of what 'British values' look like.[18]

Along similar lines, on other occasions RECs focus on specific wordings in the application that do not fit with the standard NHS REC terminology and which are interpreted as, in the words of Colin (Coastal MREC), 'vestigial Americanisms' – i.e. words and phrases typical for US-based research (normally clinical trials) which are out of place in a UK application. By and large, these should be seen as mainly an annoyance to members, but they are important in that they are seen to indicate the degree of care and attention paid by applicants; careful applicants will take the time to revise such wordings, to make them applicable to the UK. Failure to take this into consideration is held to indicate a lack of care and possibly arrogance on the part of applicants. At their most technical these issues might centre on the growth charts used to assess children: in one meeting of Northmoor LREC, Amanda asks: 'My other question which was whether US growth charts are applicable to the UK?' Megan responds 'No', with Amanda following up: 'Because the charts we use are based on an average UK weight?' Martin clarifies: 'Yes, but blood pressure chart will be calibrated for US so the weight used in the study will have to be the same.'

As noted in in the introduction, because of its status as an MREC, Coastal receives more pharmaceutical trials than the other committees and hence tends to run into more cases of 'vestigial Americanism' than St Swithin's or Northmoor LRECs. For example, one member points out that, in one application's participant information, 'There is no mention of covering expenses', but, as Rachael notes, 'They do say that "you won't be charged for taking part"!' (general laughter), with Colin addressing a pet peeve of his: 'This is an American form and no effort has been made to adapt it.' In another case, attempting to sum up the committee's views, Daniel asks, 'Are we going to want a rewrite – take out Americanisms: reference to cost of treatment, expenses for carers, results of studies should be made available to participants?' Colin 'agree[s] with Daniel. I think we ought to explain our disappointment that they sent an American PIS lightly warmed over.'

While the range of words that are regarded as problematically American are wide – 'nursing a child' rather than 'breast feeding a child'; 'IRBs' rather than 'RECs' – the key point of terminological departure, for RECs at least, is the use of the word research 'subject' instead of research 'participant'. The rise of the term 'participant' in UK medical research in the late 1990s can be traced to increased consumer perspectives and a broader drive towards patient engagement in the NHS. Yet, as Corrigan and Tutton note, 'it is unclear whether the term "participant" refers to any underlying change in research practice or in the experiences of those involved in research'.[19] This is certainly my experience in observing REC meetings where members were keen to change the terminology from 'subjects' to 'participants' with no other major alteration in terms of the interventions involved in the research (usually an industry-sponsored pharmaceutical trial) or the information given to those being recruited into the research (whatever one calls them). On the face of it, it seems odd to pay such attention to individual words without addressing other aspects of the research – involving subjects in research design, for example – which would genuinely transform them from subjects to participants. Yet within a system structured around the assessment of an applicant's trustworthiness, such apparent oddness makes more sense. As one interviewee told me:

> And you do, you just feel that if they hadn't put the effort into it they're not serious. And I make a big thing about whether they call the participants 'participants' or 'subjects'. And again, they were sort of [sighs] but, you know, I do feel that if, in their mindset, they had to remember it's a participant and not a subject, that that does have an effect, maybe not immediately but over time, *on their kind of attitude to the person taking part in the study.*

Within the context of debates around these performative aspects of 'British values', it is important to note that these reviews take place within that most British of institutions, the NHS. And as such, it is unsurprising that this institution, what the Conservative politician Nigel Lawson once called 'the closest thing the English have to a religion', should help shape REC decision making, not just in terms of the rules and guidelines for practice but also in terms of the values that underpin what members think are important. As long-term NHS watcher Rudolph Klein notes:

> In an important sense, the NHS is therefore an anomaly, not to say an anachronism … Its overall architecture … and method of funding have remained largely unchanged in a rapidly changing society. Its popularity as an institution has remained almost undiminished even while criticism of its performance has increased … precisely because it is an anomaly: an exercise in institutionalised nostalgia. Like the continuously revived films and television series celebrating Britain's wartime achievements, the NHS seems to symbolise a simpler and a warmer world of camaraderie, solidarity and national success.[20]

And in terms of REC review, this NHS-centric position is most easily examined in contrast with commercial research, a comparison that can be seen in the way in which Coastal MREC dealt with two studies. The first is the VFORCE study using a database from medical practitioners to examine prescribing practice. Chapter 2 explores this study (and the MREC members' concerns about it) in more detail, but for current purposes this specific application involves a deviation from what had previously been agreed. In the original application, patients would be informed that their clinic was involved in the study using posters and that their data might be included in the project, but there would be no direct contact with patients. In the new application, the researchers are seeking permission to contact the patients for consent, since they want to do a cheek swab for pharmacogenetic purposes. Normally, one would not think that actively seeking consent from patients would be a stumbling block in ethics review, but in this case Leonard, the lead reviewer, suggests that this was a 'precedence setting study' since it involves re-contacting people enrolled on a database where normally the data is just used retrospectively. Another member agrees, noting that the original recruitment posters did *not* talk about these sorts of studies, where data is sent back with the patient's name on it, highlighting the way this application has undermined the committee's (already limited) trust in these applicants: 'what they're actually doing is collecting samples through the back door; this will open the flood gates'.

In comparison with this commercially funded study is an application from a longstanding, publicly funded cohort study which wants to re-contact participants to seek consent for additional research on blood samples. While there are differences between what is being asked here and the VFORCE case, there are enough similarities – the same REC reviewing applications with issues around potentially re-contacting research participants to seek consent for additional investigations – to allow useful comparison. Cohort studies, a group of projects that track patients over time, have in the words of Yvonne – the lead reviewer for this application – 'produced fantastic rich data'. A few years previously, the MREC gave this particular study approval for further work which has moved on to the collection of blood samples for a range of tests, for which participants gave a broad consent. The study team promised that for every piece of research they would seek ethical approval from a REC, which was feasible in the first few years because samples were used in a 'case control' fashion – i.e. with specific individual projects, run by external researchers who accessed the cohort samples and ran an individual test. But now the size of the sample means that they can carry out more speculative research within the cohort, and the researcher does not want to have to continually return to a REC for approval for each research project. The question the REC is pondering is: 'Can they just have a "class approval" for kinds of research?', a potential problem since class approval might be going against what the original participants signed up for, thus raising the question of whether they need to be re-contacted for additional consent – a process that would be outside what the participants had originally agreed to.

Caroline, the MREC's administrator, gives some background about other committees' practices, pointing out that the MREC at Trent do a lot of this kind of work and they have a 'fast track' system, using a sub-committee and a cut-down application form which has been approved by COREC.

Simon: 'It would be grossly unethical to stop this research – it's a treasure.'
Caroline: 'It's not about stopping ...'
Malcolm: '... it's about facilitating and making legal. We put the problem to COREC and let them deal with it.'
Timothy: 'I think if we fast track, then we need to distinguish between internal analysis and data given to outside groups who will need the full REC works.'
Malcolm: 'We've identified a way in which it might be done. Caroline is aware of the Trent MREC practice and we might get approval for a similar thing from COREC.'

The technical similarities and differences in decisions between these two applications were not lost on at least one member of Coastal MREC, Barbara, who, while discussing the committee's scepticism towards the VFORCE research noted: 'The other thing is, it was a commercial organisation, wasn't it? And I think that that had a massive, massive amount to do with it.' Continuing this line of thought, Barbara then referred to the cohort study as a comparison:

> Because funnily enough, we had a professor, an academic who's got a big epidemiology study ... You know, much the same thing. To be honest with you [how it was dealt with] was vastly different. All *he* does now is put in line listings of the studies he's done in the previous months. ... And it is definitely the commercial aspect where we have

problems ... Yes, I think we [i.e. the REC] probably see the NHS as our baby, and I think we probably see it as the NHS being sort of duped into sort of you know giving away all this data that the patients don't know they're giving away, and they're being sold by this commercial company. You know, that is the big bad ... although my committee aren't into the 'big bad pharma' type thing, I know some committees are quite like that, but we tend not to be, and we have good relationships with a lot of the big pharma companies ... But I suppose it is professional trust ... It's a bit of a hard one to explain, really. But I think it is the commercial aspect of it. They [i.e. VFORCE] were selling the data for quite a lot of money to people, and the committee was saying, well, you know, you're getting this data, and the patients don't even know that you're using their data. You know, so yes, there were problems with that one.

This acute piece of self-analysis highlights the way in which REC members' decisions, embedded in the context of the NHS, draw on concerns around the correct use of health service resources and subsequent scepticism towards commercially funded research.

In terms of the protection of NHS resources RECs are concerned, most obviously, with ensuring that all costs associated with commercially sponsored pharmaceutical trials are covered by the sponsoring company. For example, in one application presented to Coastal MREC, the first reviewer is clearly unhappy with the role of the sponsoring company in the proposed research, suggesting that offering to provide the drugs required for a trial (but not actually funding it overall) is a standard 'trick' of this particular firm, arguing that in her hospital Trust, she would have it written into the contract that, since this company is not actually funding the trial, they will not get access to the data from the trial. This member wants the company's name *off* the PIS: 'this is a way for them to cheaply fund research', i.e. to fund only the drugs for 20 patients and let the NHS pick up the costs of monitoring and so on. In another case, Northmoor LREC discusses an unfunded application which would involve the use of a magnetic resonance imaging (MRI) scanner:

Megan: 'They're not seeking funding but they are using MRI scanners which we know are expensive and there's a backlog. What about the costs of equipment, even if the radiologists agree to do it in their own time? I don't think it's fair to use NHS resources like this.'

Amanda: 'I don't have a problem with the use of NHS re-
        sources if the research is done properly.'
Frank: 'Surely the first question is one of research quality
        and only when that has been satisfied can we address
        the resource issue.'
Megan: 'But in the past we have raised questions over
        interviewing staff on NHS time and undermining
        resources.'
Frank: 'This is not something we would accept.'
Megan: 'If I wanted to use NHS facilities then I would have
        to get approval and funding.'

Echoing these concerns, St Swithin's REC discusses an application that proposes to carry out 400 MRI scans, with Roderick
pointing out that this would take resources away from NHS patients. Howard, the Chair, queries whether the REC should concern
itself with resource allocation issues: 'To what extent is this our business?' Both Abigail and Ingrid reply: 'This *is* our business.' Howard
responds: 'OK, I'll ask them about the use of scarce resources.'[21]

At this point, as an interim summary, it is clear that REC members read application forms for a number of reasons; most obviously, to gain an understanding of what a researcher wants to
do. Yet these forms also provide raw material for an assessment
that goes beyond just an understanding of how a particular research project is going to be carried out, an assessment that is
more about aspects of an applicant's character and whether they
can be trusted to do what they say they are going to do. In this
context, of the 'virtual witnessing' of researcher character and behaviour, the PIS is a vital part of the application, not least because
it gives direct evidence about what the researcher says they are
going to tell people recruited into the project. Since one of a REC's
core roles is to ensure that the information presented to people
who might take part in research is accurate and understandable, a
great deal of a committee's time is spent reviewing and correcting
such sheets.[22]

From a contemporary point of view, debates about informed
consent before the development of RECs are really quite peculiar
in that the core aspects of modern discussion (the wording and
length of the participant information sheet, the nature of the consent form) are entirely absent. Indeed, not only was written and

signed consent not required, in the early 1960s the official position was to *discourage* it. This point of view is most obviously stated in the report 'Responsibility in Investigations on Human Subjects' by the MRC (roughly the British equivalent to the US National Institutes of Health), published as part of its 1964 annual report. Here the MRC starts by suggesting 'That it is both considerate and prudent to obtain the patient's agreement before using a novel procedure is no more than a requirement of good medical practice' and that although occasionally patients should not be told they are being enrolled into 'controlled trials', 'In general ... patients participating in them should be told frankly that two different procedures are being assessed and their co-operation invited.' The MRC is quite clear as to what counts as best practice around seeking such consent – which should be 'freely given with proper understanding of the nature and consequences of what is proposed'; it should be oral consent which 'the investigator should obtain ... himself in the presence of another person'. That this consent is oral is explicit. Indeed, researchers are warned that 'Written consent unaccompanied by other evidence that an explanation has been given, understood and accepted is of little value.'[23]

That what the MRC was setting out here was simply the understanding of current practice is borne out by an interview with a medic who was involved in research at the time, who noted that ethical debate about research 'wasn't a huge problem at that time because everybody thought that the work they were doing in clinical research in this country, both therapeutic and non-therapeutic was in fact governed by consent but it was entirely *oral* consent'. This interviewee recalls: 'certainly people going to see a patient and saying "look, we're just going to do a little test" and the test might be a liver biopsy which carries mortality risk'. The position of the time was set out by John McMichael, the Chief of Medicine at the Hammersmith Hospital, who this interviewee quotes as saying:

> 'that the responsibility for the experiment lay with the investigator and it must be the investigator who decided what was done, not the subject'. Now that was very much in keeping with the paternalist attitudes of medicine at that time because you didn't explain to somebody who got cancer. You used some euphemism.

Talking about his own work, he remembered

> doing experiments at that time using radioactive material but didn't
> explain it was radioactive material to the patients at all. That was an
> era when the whole world was looking for peaceful uses for, atomic
> energy and so we had no reason to explain what we were doing and
> very few explanations were given at that time in respect of, of research.

Against this background, it is unsurprising that the initial 1973
guidance for RECs, drafted as it was by the RCP, held the same
position as the MRC.[24]

However, by the mid-1970s, it is clear that the 'witnessed oral
consent to research' model was under strain, especially from changes
in clinical practice. In 1968, guidance from the Medical Defence
Union to doctors (especially surgeons) seeking to avoid legal action
when treating patients advised that 'written consent is preferable
to an oral consent because it can more easily be proved to have
been given', especially in the case of 'any operation which requires
a general anaesthetic or to any procedure which may involve a special risk'.[25] In this context, the arrival of RECs and the anticipatory
nature of their review added to the pressure on witnessed oral consent. Indeed, for L.J. Witts – Emeritus Nuffield Professor of Clinical
Medicine at Oxford – the main purpose of an ethics committees
was, in 1973, 'to give meaning to that phrase "informed consent";
we know that the average patient can never really understand
exactly what is involved but he likes and trusts his doctor and tells
him to go ahead. The committee makes sure that the procedure does
satisfy certain criteria.'[26] And, of course, the prior nature of ethics
review requires written consent materials for the REC to review.[27]

Building on the role of written consent as a defence against clinical malpractice suits, the second way in which the arrival of RECs
pushed informed consent in a written direction centres on concerns
around the legal liabilities associated with different kinds of consent. That RECs should protect British medical culture from the
vagaries of litigation was clearly the suggestion of at least one con-
tributor to the *British Medical Journal* who hoped that

> the ethical committee should aim to spread its ideas throughout the
> whole institution … unless we do create this climate doctors will be
> pushed into a medicolegal position rather than an ethical one, as

already occurs in the U.S.A. – where the patients have to sign a long involved form containing a lot of small print.[28]

However, his hope that RECs would save doctors from 'long involved form[s] containing a lot of small print' was misplaced; indeed it was the RECs themselves in the appointment of lay people – often lawyers or even judges – that opened up biomedical research practice to legal scrutiny in a way that had not happened before, drawing attention to the issue of who gets to sign off on the informed consent – the researcher/witness or the participant.[29]

These issues – partly procedural, partly legal – can be seen in a discussion held in an RCP-convened meeting of 'Chairmen of Ethical Committees', in 1976. The discussion starts with Drs Edwards and Nordin (Chairs of RECs at St Mary's Hospital and Leeds, respectively) asking 'whether or not it was legally advisable to have a consent form signed by a patient who had verbally agreed to take part in a research project'. Sir Cyril Clarke, President of the College and Chair of this meeting, responds by quoting from the 1973 report (see above) and confirming that the two signatures required should be from the witness and the person giving the explanation (i.e. *not* the patient), arguing that if the patient 'did sign the research worker could be in a weaker position if anything went wrong, because the relatives could say that the subject had done so without understanding what he had been signing'. At this point, Mr C.H.H. Butcher, a solicitor involved in Medical Defence Union work and an advisor to the committee, questions the sense of the 'non-signing arguments', pointing out the very limited number of legal claims arising out of medical research and therefore the limited risks of doctors being sued. Dr Booth (from Hammersmith Hospital) 'supported Mr. Butcher on this point', highlighting the issue of anticipatory review: 'one of the functions of an ethical committee should be to check that the form represented what was, in their opinion, "informed consent"'. A number of other Chairs then point out that – often as a result of legally qualified lay members – their committees *did* require written consent and signature. For example,

the ethical committee at King's College Hospital ... had taken advice from colleagues, and their lay member, who was a member of Queen's Counsel [i.e. a senior lawyer]. The latter's advice had been that in a Court of Law, if there was negligence, it may be that the

Judge would take a more lenient view if everyone, the subject, the
investigator and the witness, had signed.

Similarly, 'At St. Thomas' no investigations could be undertaken
unless the form was signed by the patient or subject concerned.
This had been stipulated by their lay member, a judge' and 'this
was also a requirement in Bristol, which had taken legal advice'.[30]
Indeed, one of my interviewees who was a member of one of the
first RECs noted:

> ours [REC] was set up quite early on at [hospital name] and I think
> it functioned quite well. The major question we came across was
> the consent forms because we moved into the question of having
> written, signed, consent forms quite quickly and one of the features
> I remember so well for lay members of committees of that sort, was
> they were very good at advising on consent forms, which would be
> understood by lay people.

Just over two years later, similar concerns arose at another Chairs
of Ethical Committees meeting. Dr Denham, of Northwick Park
Hospital, pushed back against the trends in REC practice, saying:

> that although local opinion favoured written consent, his ethical
> committee took the view that normally verbal consent was sufficient
> ... written consent was rather a formal procedure, and did not neces-
> sarily guarantee that an explanation had been given; patients might
> feel, quite wrongly, that by giving written consent they had waived
> any opportunity to withdraw or complain; also they might argue at
> a later date that they had not understood what they were signing.

In response, the President of the College pointed out that, 'following a recommendation in the first report of the Health Service
Commissioner' the 'DHSS [Department of Health and Social
Security] had devised a form to record a patient's consent to participate in clinical research investigations ... and had sent [it] to
the College for comments'. There then follows an extended discussion over the problems associated with such a standardised form,
mainly focused on the loss of flexibility that such standardisation
would bring. Even Mr Butcher, the solicitor in favour of written
consent in general, felt that the standardised form 'would be too in-
flexible to cover the points which a judge would take into account
when deciding whether or not a true consent had been given'.[31]

While such a standardised consent form for research in the NHS did not seem to have progressed beyond the consultation stage, the idea of written consent did make its way into the next set of guidance for RECs from the RCP which notes that: 'Obtaining true (or informed or understanding) consent is central to the ethical conduct of clinical investigations' and that while

> trivial or minimal risk procedures ... may be done with a simple verbal explanation and verbal response ... more substantial procedures should be the subject of an explanatory document setting out the purpose of the investigations ... The subject may study this and then sign a paper that states that the document has been studied and discussed with the investigator and that the subject agrees to participate.

While witnessed consent is not rejected, it is no longer presented as the 'default' setting but rather being 'especially useful in the old and those who have intellectual or cultural difficulties in speech or understanding.'[32] Highlighting the apparent spread of written consent, one contemporary critic noted that: 'Most, although not all, research ethics committees insist that every subject taking part in research projects gives informed consent to the research in writing.'[33]

While the need for *written* informed consent was putatively based on legal concerns about researchers' liability (often on the part of legally qualified lay REC members), the actual risks of not asking participants to sign forms remained uncertain. In 1984, even with regard to clinical practice, the legal status of informed consent (in whatever form) was up for debate, with the legal correspondent of the *BMJ* confidently dismissing 'informed consent' as a US 'doctrine ... founded on the proposition that it is for the patient, not the doctor, to determine what happens to a patient's body' which 'has so burdened the profession with costly litigation that in some states legislation has been introduced to modify' it. Happily (for the author) 'the English courts are certainly not going to accept it as part of the law of England'.[34] A similar tone – certainly with regard to the perceived American roots (and hence inappropriate nature) of informed consent – can be found in a discussion during a meeting of the RCP's Committee on Ethical Issues in Medicine in 1986 where informed consent was dismissed as 'really an American concept implying a knowledgeable patient rather than a

knowledgeable doctor and was accompanied by much documentation. English law dealt much more with the reasonable doctor than
the reasonable patient.' The response to this, from the legal phil-
osopher Gerald Dworkin, seems, from a contemporary position,
hard to argue with: 'The important issue was the patient's right
for information rather than the doctor's judgment as to how much
to give.'[35] As spelled out in a letter to the *BMJ* in 1988, it is not
the case that 'the signed consent form fulfils a legal requirement'
but rather that, given that 'Before any examination, procedure, or
surgical operation the doctor requires the informed consent of the
patient or he may be guilty of an assault', in some cases, rather than
oral consent:

> it may be prudent for the doctor to have some record of having
> obtained the patient's informed consent to help him defend a subse-
> quent action by a patient claiming that such consent was not given.
> This is the purpose of the consent form. It is not a legal requirement,
> since a doctor may well be able to prove by other means – for in-
> stance, by the evidence of witnesses – that the patient did understand
> the nature and effects of surgery and agreed to them.[36]

Even in 1989 debates around the RCP's next revision of its guidelines for RECs, a strong case was made against written consent:

> Although it was known that various patient organisations were
> strongly in favour of written consent, concern was expressed that
> this procedure might somewhat unethically imply a desire to pro-
> vide a safeguard for the research worker, upon whom the obligation
> rested to ensure that all procedures had been correctly followed.

However, given that 'the Research in Patients working party
[drafting the guidelines] had come down in favour of provision
of a clearly set-out document which asked a series of questions
requiring satisfactory answers before signature', the resistance to
written consent seemed to be on its last legs.[37]

Similar discussions around a draft of the Department of Health's
1991 'Red Book' – the document setting out formal requirements
for LRECs for the first time – noted that 'It was felt that the necessity for a witness to attest to consent at all times should be recon-
sidered; the ethics committee should consider whether a witness
was in fact appropriate.'[38] The Red Book itself makes a clear case

for written consent: while 'The procedure for obtaining consent will vary according to the nature of each research proposal ... The LREC will want to be satisfied on the level and amount of information to be given to a prospective subject' and 'Written consent should be required for all research (except where the most trivial of procedures is concerned).' The College's revised guidelines themselves mirror the 1984 (first) edition, with written consent required for 'all but the most trivial procedures' in research with healthy volunteers and patients involved in non-therapeutic research.[39]

Thus written PIS and signed consent forms provide the limits of the kinds of trust decisions that RECs can make. Perhaps if they were even more rooted in professional self-regulation – with no non-medical members, for example – then RECs would be able to take applicants at their word that they would discuss the proposed research and seek their consent. This is, after all, the case in the early days of IRB review in the US, where prior review by colleagues was used as a way to *avoid* the need for written consent documents. Yet in the UK, non-medical lay members were appointed to RECs very early in their development, and organisational changes meant that many lay members were drawn from a pool of people who instinctively engaged with the medical profession in a critical manner.[40] There was thus a need for a REC to ensure that, as Chris Booth from Hammersmith REC put it in 1976, 'the form represented what was, in their opinion, "informed consent"'.

Of course, this model of consent still requires considerable trust in the researcher involved. The REC has no way of knowing whether the information materials approved in the application are the ones that will *actually* be used in the clinic. The REC has no way of knowing whether the consent forms will *actually* be signed. Yet beyond the crucial role of RECs in the development of written consent, such materials provide a crucial way for REC members to assess applicants' trustworthiness.[41] At the most basic level, such assessment is around what Laura Stark calls 'housekeeping work', which focuses on issues such as spelling mistakes and other, apparently minor, stylistic errors, with IRB 'members us[ing] the apparent degree of care taken in submitting a tidy application as a proxy for an investigator's self-discipline and fastidiousness'.[42]

For NHS RECs, these character-based readings of attention to detail happen but do not necessarily play a central role in assessing an application. In Coastal MREC, Adam admits to being irritated and complains about 'a thinness to it that comes out when you look at the PIS. One is full of typos, and jargon. It does not say anything about where interviews will happen, confidentiality etc.' There are similar issues in the other PIS. In St Swithin's LREC, Howard makes a wry aside: 'it's astonishing that with modern technology, including things called "spell checkers", it's amazing that people do not spell check their PIS'.

While such judgements do not dominate REC assessment of consent materials, as an indicator of the way in which RECs are pulled towards discussion of competence and trustworthiness, it is worth noting how even concern of an applicant's *potential* mis-understanding of a REC's position can trigger trust concerns. For example, St Swithin's REC discusses a project requiring recruitment of parents from several different groups:

Edward: 'Do we want three different [PIS] forms [for the parents in each group]. Isn't that a waste of paper?'

Hilary: 'I think it needs to be clear, they do need separate sheets [for each group].'

Howard: 'How about three bullet points and you cross out the irrelevant ones?'

Hilary: 'Just as a point of principle we do ask people to be invited for a specific reasons ...'

Ben: '... to personalise it, so people are not seen as research fodder.'

Hilary: 'Will researchers really cross things off or tick boxes?'

Howard: 'Well, would they print off the right sheet?'

Hilary: 'Well, I would, but I know people who wouldn't.'

Howard: 'There's no point us requiring complicated things if they cannot do it.'

Ben: 'This is the simplest thing. If they can't get the sheet right then they shouldn't do the study!'

Here, concerns about the resource implications of requiring different PIS for different groups shade across into the need to treat participants with respect – by avoiding the lack of personalisation and the impression of 'research fodder' given by crossed-out bullet points – before basing the need for three different PIS on a basic requirement for competence and future behaviour ('If they can't get the sheet right then they shouldn't do the study!') via the question of trust ('Well I would, but I know people who wouldn't').

But at the same time, members make a number of critiques of consent materials that, while important in terms of the success (or not) of an application, are not necessarily to do with trustworthiness. The most obvious cases are when RECs express concerns about the perceived 'coercive' nature of the wording in a PIS. While REC members are clearly concerned by such wording, these concerns are not couched in terms of indicating a researcher's untrustworthiness, but rather in curt instructions for revision. A discussion in Coastal MREC turns to the PIS which 'does need better formatting' and an introductory letter which 'talks about "this important study" and whenever I hear the word "important" I think "coercive"'; in another application there is unhappiness with the coercive tone of the word 'important' – 'it is important that you keep your appointments for the study to be worthwhile'. Similar focus on the tone of specific words can be seen in a discussion in St Swithin's LREC, with Tom suggesting that the language used was far too emotional (talking about 'allowing tissue to perish' or 'any refusal to give permission').

When RECs do interpret PIS and other consent materials in terms of an applicant's trustworthiness, they tend to base their character assessments around whether such materials are 'misleading' or not. In Northmoor LREC, Neil begins his discussion of a particular PIS: 'I think they are underplaying the blood samples. There is no mention in the PIS and the cartoon [attached to the PIS for the children] is misleading.'

Martin: 'Yes, there's no needle in the picture! And I wouldn't mind doing injections if all the children did was smile.'

Neil: 'The application does say that "a local numbing agent is allowed" – I think we should ask them to use this automatically.'

Martin: 'The blood test is mentioned in the PIS actually ...'

Neil: 'But they need to mention it in the 'discomfort' section of the statement.'

Megan: 'But local anaesthetic is mentioned ...'

While there is some ambivalence here, it is clear that glossing over the role of blood samples – 'underplaying' – in the PIS is not viewed favourably; indeed it might be seen to 'colour' the assessment of even those parts of the document where the approach adopted might be regarded as correct (including an anaesthetic). Leaving out information seen as important by the REC is a common trigger for committee members to read a PIS as misleading, as can be seen from an application before Coastal MREC, where the comment is made regarding PIS, of which there are two, one for children (10–18 years) and one for adults: 'I found the one for children more confusing.' The PIS are not good; they are unclear and they mislead over the unlicensed nature of the drug (they describe it as experimental) and how many people have previously taken the drug (the wording suggests that over 100 people have taken the drug 400 times each but actually only 400 doses have been prescribed in total).

As this latter case suggests, a key way for REC members to decide whether an application is misleading or not is whether there are internal inconsistencies between the various documents that make up an application. Is there a discrepancy between the application form, the investigator's brochure, and the PIS regarding how many people have previously taken the drug? Are participants told a drug is 'safe' in the PIS when the application form acknowledges some risks associated with its use? In one case, St Swithin's move on to a PIS, which Dean feels is misleading; the application points out that material administered to subjects carries the risk of an allergic reaction, while the PIS describes the procedure as 'extremely safe'. It was agreed that the word *extremely* needed to be struck from the PIS and the applicant encouraged not to use it in the future. Abigail summed up: 'Don't use "extremely"; it's a bit like the F-word. If you haven't got a better vocabulary then it really is a pity.'

In a similar example, a discussion in Coastal MREC notes that the answers to two questions on the application form contradict one another over whether there will be a signed record of consent. The application is described as a bit misleading, underlining the view that these problems are minor points, although they don't convey the idea of someone who has thought through their research properly. Likewise, in a meeting of St Swithin's LREC, Susan is concerned about the DNA sampling sheet in an application, which has two parts, one of which refers to a 10 ml sample at the baseline visit and a statement that the clinical data could be fed back to the patient. But the application itself suggests that the samples will be numbered and no feedback will be possible. Thus for her, 'the PIS is unclear on this and may give a false impression to participants'.

Thus it is easy to see how RECs come to interpret such intra-application inconsistency in moral terms, as with the application reviewed by St Swithin's, where Tom (expert) points that that there is no mention in most of the paperwork of the establishment of permanent cell lines from sampled material, yet it is clear that hidden away in the form there is a suggestion that this might happen. Tom did not have any objection to the establishment of such lines, but the possibility is not made clear to participants, who are told in the PIS that their tissue will die and not be used for therapeutic purposes.

The 'virtual witnessing' offered by consent materials gives REC members a sense of insight not just into a researcher's current trustworthiness, but also into future behaviour. For example, Northmoor and District LREC turns to a study sampling human stem cells from bone marrow. Frank, a lay Vice Chair, provides the lead review, noting that what the researchers want to do is, 'entirely straightforward', in that they want to use discarded patient tissue (from surgery) to see if they can develop stem cells. With regard to patient consent, 'They say that patients at the hospital sign away tissue for research, but it is a shame they haven't sent a consent form to us.' Martin, the committee's Chair, responds that 'it [the form] *was* seen by a previous ethics committee but it's reasonable to ask to see it'.

The applicant is invited in and begins to explain that the application is for broad approval of tissue sampling, prior to separate applications for specific individual research projects, stating that 'we are trying to ask for continuation of the early stage work we do which then leads on to specific studies, for which we ask separate ethics approval'. Martin then asks 'if we could have a look at the consent form that you use?', at which point the applicant fishes one out of their bag. The consent for research form is a section on the back of the surgical consent form, offering 'opt-out' consent from research use of discarded tissue. As it is passed round the committee, discussion continues.

Applicant: 'This is a general operative form for the hospital.'
Martin: 'Has this been revised in the past few years – since most hospitals have changed their forms over the past few years?'

Applicant: 'I don't know about the whole thing, but certainly
the section approved by the research ethics committee hasn't changed.'

The broader committee then begins to ask questions, focusing
on the 'opt-out' nature of the consent being sought:

Frank: 'Don't you think that's unethical?'
Applicant: 'No, they [the patients] are taken through it and
some people do opt out. We know that it works.
We don't let the scientific team turn up before the
operation, which is coercive. The Helsinki agreement means that the surgeons have to do the
consent, not us.'
Martin: 'Standards of ethical consent have changed since this
form was approved.'

The applicant points out that the hospital has printed many
copies of the form and would not be happy if they had to produce
new ones. The applicant leaves.

Megan is unhappy with the consent form, articulating her concerns in clear terms of distrust: 'I don't believe [the applicant's] reassurance that the patient is taken through the form.'

Martin: 'The problem we have, is that it has been approved
by another committee ...'
Megan: 'But that may be OK in the past, but nowadays that
would not stand up. In maternity we have to [i.e.
full opt-in] consent people to use placenta material
but many people don't consent.'
Frank: 'But the placenta is different isn't it?'
Megan: 'How?'
Rebecca: 'It's a very emotional issue.'

Rebecca then tries to reassure Megan in terms of expectations around this hospital's trustworthiness: 'But the hospital
has a history and research reputation' to which Megan responds:
'But we can't base our decision on practice and culture ...', essentially requiring a trust decision to be made anew for each
application.

In his role as Chair, Martin tries to clarify the concerns:

There are two issues. First is the issue that this is not consent for a specific project and also it's [i.e. the form] been approved by another committee. It ought to be possible to give a provisional opinion, part of which requires changes to the consent form. We could take the position that they need to separate research consent off from surgical consent.

Frank's position is that the current form provides enough information for people to understand what it is they are consenting to: 'I really don't know what we're arguing about. It's clear on the consent form.'

Martin: 'Are people happy with the generic consent? Are there specific things that the committee feels they should do to change the form?'

Sally: 'It doesn't say "if you don't want this, cross this through".'

Megan: 'They're asking that people agree. Most people will think it's a consent to surgery.'

Martin: 'This is the same as normal studies.'

Megan: 'But it's not, since in a research study [with 'opt-in' consent] you know that's what it's for.'

Martin: 'We could add, visibly, "If you do not want your tissue to be used in this way, please cross it out". We could ask for a rewrite but that would be a significant change.'

Megan: 'I think it's sneaky, a sneaky way of getting access to lots of tissue without telling people too much.'

Frank: 'I disagree, it's clearly stated.'

Martin sums up, trying to present a compromise position: 'OK, there needs to be a written statement about crossing out. *We are not just relying on someone to tell the patient.* It needs to be clear that people can opt out and the patient needs a record.'

There is a clear lack of consensus in this case, between Frank's rather phlegmatic position that the current opt-out consent is clear enough, and Megan's belief that not only is the consent section likely to mislead patients, but that this is an intentional feature of the consent form, raising questions of character about the researchers, or at least the original drafters of the surgical consent form. While this case – with Martin's clear statement of distrust: 'We are not just relying on someone to tell the patient' – is at one end of the spectrum, consent materials, forms, and PIS are crucial in allowing RECs to virtually witness researchers' future behaviour.

This chapter has set out the various ways in which RECs can interpret ethics applications (and associated paperwork) in terms of applicants' trustworthiness. For REC members, application forms, PIS and other written material provide crucial insight not just into what researchers *say* they are going to do, but what they are *actually* going to do, as well as their underlying character. This virtual witnessing of researchers' future behaviour and character allows RECs to make trust decisions. Of course, given the peculiar asymmetries that characterise the nature of these trust decisions, it is important to note that RECs have little or no way of validating the accuracy of the trust warrants they employ; the nature of the process means that only rarely do RECs discover whether, for example, researchers have altered consent materials without approval.

Yet written applications are only one aspect – albeit the most visible – of information available to REC members in their assessment of applicants' trustworthiness. Such judgements are made within the context of pre-existing relationships – either through regular applications to a REC or working alongside members in the clinic – and a process that encourages applicants to attend meetings and answer questions about their research face to face with RECs. The next two chapters address these features of trust decision making in turn.

Notes

1 *The Association of Research Ethics Committees Newsletter* 5 (2001), 8.
2 'Reading trust between the lines: "Housekeeping work" and inequality in human-subjects review', *Cambridge Quarterly of Healthcare Ethics,* 22 (2013), 394.
3 Alison While, 'Research ethics committee at work: The experience of one multi-location study', *Journal of Medical Ethics,* 22 (1996), 352–355; Konrad Jamrozik, 'Research ethics paperwork: What is the plot we seem to have lost?', *British Medical Journal* 329 (2005), 286–287.
4 Steven Shapin, 'Pump and circumstance: Robert Boyle's literary technology', *Social Studies of Science,* 14:4 (1984), 484 and 491.
5 Arthur Caplan, 'Random-sampling: A modest proposal for reforming IRB review', *IRB: Ethics & Human Research,* 4:6 (1982), 8.

6 Carol Heimer and J. Lynn Gazley, 'Performing regulation: Transcending regulatory ritualism in HIV clinics', *Law & Society Review*, 46:4 (2012), 881.

7 The meetings I observed took place under the third version of the SOPs, dated June 2005 (COREC, *Standard Operating Procedures for Research Ethics Committees*, Version 3.0, June 2005 [NPSA]), a 208-page document setting out everything from a glossary of terms to the processes involved in applying for REC approval, to the format for the agenda for REC meetings.

8 The disproportionate influence on decision making of initially presented information – the 'anchoring effect' – is a well-established feature of individual and group decision making: for a review, see: Adrian Furnham and Hua Chu Boo, 'A literature review of the anchoring effect', *The Journal of Socio-Economics*, 40:1 (2011), 35–42. But, as this book makes clear, REC decisions are not pre-determined when the committee sits down to discuss an application. The input from other members of the REC regarding, for example, their personal knowledge of the researcher (Chapter 2) or their own scientific expertise (Chapter 4) or the performance of an applicant in front of a committee (Chapter 3) can all radically re-shape the direction of a REC's decision making.

9 The prevalence of the 'provisional' decision can be seen in the following aggregate data, taken from annual reports, for all three committees over a 3-year period (2004/2005 to 2006/2007): Northmoor, 59%; St Swithin's, 65%; Coastal, 75%.

10 Similar accounts of the importance of physically meeting for REC decisions can be found in Sarah Dyer's work, where she notes the chair of one REC saying: 'the fact that the committee actually meets to discuss it is a great strength because often we start to bounce ideas off each other and we realize that there is an elephant in the room and no one has realized': Sarah Dyer, 'Applying Bioethics: Local Research Ethics Committees and their Ethical Regulation of Medical Research' (PhD Dissertation, University of London, 2005), 90.

11 West Birmingham Community Health Council, *Ethical Committees: A Seminar for CHC Members on District Ethical Committees held on 21st April 1988* (Birmingham: West Birmingham CHC, 1989).

12 Royal College of Physicians, *Guidelines on the Practice of Ethics Committees in Medical Research Involving Human Subjects* (London: Royal College of Physicians of London, 1984/1990), p. 10 of the first edition and p. 11 of the second; The Department of Health, *Local Research Ethics Committees* (The 'Red Book') (London: HM Stationery Office, 1991), 9; Leigh & Baron Consulting and Christie Associates, *Standards for Local Research Ethics Committees*.

A Framework for Ethical Review (Bristol: McKenna, NHS Department of Health Training Committee, 1994), 9; Donald Evans, Martyn Evans, David Greaves, and Derek Morgan, *Trainer's Manual: For the Training of Members of Research Ethics Committees* (Swansea: Centre for Philosophy and Health Care, University College, 1992), sections 3.5–3.6.

13 With regard to the importance of clarity in trust negotiations, Susan Shapiro notes, 'Norms about disclosure apply particularly to agents in information dissemination and interpretation roles, for example, journalists, accountants, or scientists. Procedural regulations embellish the norm. They touch issues such as standards of proof (corroboration, replication, or second opinions), sampling, randomization, surprise or spontaneity, control groups, statistical inference (or other assessments of validity, reliability, alternative interpretations, side effects), confidentiality or the proprietary nature of information, and the threshold (what is material, what can be omitted) and timing of disclosure': Susan Shapiro, 'The social control of impersonal trust', *American Journal of Sociology,* 93:3 (1987), 637.

14 Within the NHS, certain activities – clinical audit and service evaluation – which resemble research are exempt from REC review since they are not seen to generate 'generalisable knowledge'. For a discussion of some of the ethical problems associated with this (possibly arbitrary) distinction, see: Derick Wade, 'Ethics, audit, and research: All shades of grey', *British Medical Journal,* 330 (2005), 468–471.

15 This echoes a key insight from Laura Stark's work on US IRBs, namely that IRB 'members use the documents that researchers send to the board to judge the character of the researchers. For board members, the style and tidiness of researchers' documents offer a snapshot of the person behind the application': Laura Stark, *Behind Closed Doors,* 15–16.

16 Carol Heimer and J. Lynn Gazley, 'Performing Regulation', 880.

17 For a discussion of issues around the implications of monolinguism in ethics review, see: Laura Stark, 'The Language of Ethics: How Ethics Review Creates Inequalities for Language Minorities in Research', in Will van den Hoonaard and Anne Hamilton (eds.), *The Ethics Rupture: Exploring Alternatives to Formal Research-Ethics Review* (Toronto: University of Toronto Press, 2016), 91–103.

18 In terms of ethnicity, monitoring data puts 89% of lay members and 91% of professional members as describing their ethnic origin as white, which is in line with the overall population reported in the 2001 National Census, i.e. 92%. In terms of education (which we might use as a proxy for class), the high levels we might expect for expert members are matched by those for lay members, of whom 84% were educated to degree level or above (including 50% educated to

postgraduate level), in noticeable contrast to the 20% of the population aged 16–74 in England and Wales educated to degree level or
above: Lucy Simons, Gill Wren, and Sarah Buckland, *Survey of Lay
Members of Research Ethics Committees (RECs)*, Report for Involve
(Hampshire: National Institute for Health Research, 2009).

19 Oonagh Corrigan and Richard Tutton, 'What's in a name? Subjects, volunteers, participants and activists in clinical research', *Clinical Ethics,*
1 (2006), 102. For further debate, see: Petra Boynton, 'Letter: People
should participate in, not be subjects of, research', *British Medical
Journal,* 317:7171 (1998), 1521; Iain Chalmers, 'Letter: People
are "participants" in research', *British Medical Journal,* 318:7191
(1999), 1141.

20 Nigel Lawson, *The View from No. 11: Memoirs of a Tory Radical*
(London: Bantam, 1992), 613; Rudolph Klein, *The New Politics of
the National Health Service* (Harlow: Prentice Hall, 2001, fourth
edition), vii.

21 As Chris Goldsworthy pointed out to me, there is double meaning
here, based on many REC members being employed by the NHS and
thus acutely aware of the burden under-funded research can present to
health service resources. This is *their* business both because they are
REC members but also because they too are employed by the NHS.

22 The '*GAfREC*' document that serves as the framework for NHS
REC practice makes the importance of reviewing consent materials
explicit: 'The primary task of an REC lies in the ethical review of
research proposals and their supporting documents, with special attention given to the nature of any intervention and its safety for participants, *to the informed consent process*, documentation, and to the
suitability and feasibility of the protocol' (emphasis added) with a 'requirement for a favourable opinion' listed as 'a full description of the
process for obtaining informed consent, including the identification of
those responsible for obtaining consent, the time-frame in which it will
occur, and the process for ensuring consent has not been withdrawn
... [as well as] ... the adequacy, completeness and understandability
of written and oral information to be given to the research participants': Department of Health, *Governance Arrangements for NHS
Research Ethics Committees* (London: Department of Health, 2001),
23 and 26.

23 Medical Research Council, 'Responsibility in Investigations on Human
Subjects', in *Report of the Medical Research Council for the Year 1962–
1963* (London: HMSO, 1964), 22–23. The report was considered so
important that it was reprinted in the *British Medical Journal* later in
the same year (1964; 2:178–180) to extend its audience. For more information on the MRC, see: Carsten Timmermann, 'Clinical Research

in Post-War Britain: The Role of the Medical Research Council', in Caroline Hannaway (ed.), *Biomedicine in the 20th Century: Practices, Policies and Politics* (Amsterdam: IOS Press, 2008), 231–254.

24 Royal College of Physicians of London, *Report of the Committee on the Supervision of the Ethics of Clinical Research Investigations in Institutions* (London: Royal College of Physicians of London, 1973), 1.

25 Medical Defence Union, *Consent to Treatment* (London: Medical Defence Union, 1968, revised edition), 3. The Medical Defence Union was the world's first medical defence organisation, providing legal advice to doctors for incidents arising from their clinical care of patients.

26 M.D. Eilenberg, Roger Williams, and L.J. Witts, 'New horizons in medical ethics: Research investigations in adults', *British Medical Journal*, 2:5860 (1973), 222.

27 Arguments between medical researchers over the appropriateness of oral versus written consent go back at least to 1947: see Chapter 2 of Tal Bolton, 'Consent and the Construction of the Volunteer: Institutional Settings of Experimental Research on Human Beings in Britain during the Cold War' (PhD thesis, University of Kent, 2008).

28 M.D. Eilenberg et al., 'New Horizons in Medical Ethics', 222.

29 Interestingly – and a crucial reminder of the pseudo-isomorphic nature of ethics review bodies in different countries – the rise of IRB review in the US was seen as a defence *against* the need for legalistic written consent since 'group review was designed to allow researchers to avoid consent documents … The Clinical Research Committee [a proto-IRB at the National Institutes of Health] was researchers' tool of choice through which they could get formal permission *not* to use signed forms … to counteract lawyers' demands for subjects' signed forms': Laura Stark, *Behind Closed Doors*, 135.

30 RCPA MS4930/1, 'Report of a meeting of Chairmen of Ethical Committee or their representatives held on Wednesday, 2nd June 1976', 9–10 (paras 9.1.1–9.1.14).

31 RCPA MS4930/1, 'Report of a meeting of Chairmen of Ethical Committee or their representatives held on Tuesday, 26th September 1978', 3–6 (para 5.1.1–para 5.1.14).

The Health Service Commissioner (now called the Parliamentary and Health Service Ombudsman), set up in 1973, is an appeal service for complaints about the NHS that have not been resolved by other means. The report in question highlights the legal problems involved in non-written consent and investigates a complaint by a man 'about the treatment his wife had received at various hospitals prior to her death in July 1973', the first of which was that, 'following an operation in January 1969 his wife had been asked and had agreed to take part in an experiment involving tests the details of which had not been explained to her; he believed these contributed to her death'. The Commissioner found that, although 'no evidence could be found that the wife had taken part in a research project … it was beyond reasonable doubt that she had in fact taken part in tests which were being undertaken in the associated medical school at that time'. The doctor running the research, who had moved to Australia and taken all written records with him, claimed that she had given oral consent to be involved.

As a result, the Commissioner recommended that participation in a research project should be recorded in a patient's notes, with the hospital concerned 'also considering the introduction of a form of written consent to participation in a research project': Health Service Commissioner, *First Report of the Health Service Commissioner, Session 1974–1975* (London: HMSO, 1975), 50.

32 Royal College of Physicians, *Guidelines on the Practice of Ethics Committees in Medical Research* (London: Royal College of Physicians of London, 1984), 12.

33 Peter Lewis, 'The drawbacks of research ethics committees', *Journal of Medical Ethics*, 8 (1982), 62.

34 Anon (Our Legal Correspondent), 'News and notes: Medicolegal – what should a doctor tell?', *British Medical Journal* 289 (1984), 325.

35 RCPA MS 5062, 'A meeting of the College Committee on Ethical Issues in Medicine was held at the College on Tuesday 25th March 1986', 6.

36 Niall Warnock, 'Letter: How informed is signed consent?' *British Medical Journal*, 296 (1988), 1126.

37 RCPA MS 5062, 'A meeting of the College Committee on Ethical Issues in Medicine was held at the College on 3rd May 1989', 2.

38 RCPA MS 5062, 'A meeting of the College Committee on Ethical Issues in Medicine was held at the College on Monday, 6th November 1989', 3.

39 Department of Health, *Local Research Ethics Committees HSG(91)5* (London: HMSO, 1991), 12; Royal College of Physicians, *Guidelines on the Practice of Ethics Committees in Medical Research Involving Human Subjects* (London: Royal College of Physicians of London, 1990 (second edition), 20.

While it is not referred to in these discussions, it is probable that broader contemporaneous developments in the regulation of research further cemented the central role of written informed consent and its review by an ethics committee as a prerequisite for ethical research. In 1990 the then European Community produced its guidance on good clinical practice for pharmaceutical trials, which both required ethics review of any clinical trial of a new drug, and strongly promoted written consent materials and participant signed forms. The guidance allows signed witness consent in therapeutic research, but 'Consent must always be given by the signature of the subject in a non-therapeutic study, i.e. when there is no direct clinical benefit to the subject': CPMP Working Party on Efficacy of Medicinal Products, 'Good clinical practice for trials on medicinal products in the European Community', *Pharmacology & Toxicology*, 67:4 (1990), 365, para 1.14.

40 The first cohort of lay members tended to come from Community Health Councils (CHCs) which were bodies set up in 1974 to provide consumer representation within NHS management. For more on the experience of CHC lay members of RECs, see: West Birmingham CHC, *Ethical Committees: A Seminar for CHC Members on District Ethical Committees held on 21st April 1988* (Birmingham: West Birmingham CHC, 1989). For CHCs in general, see: Rudolph Klein and Janet Lewis, *The Politics of Consumer Representation: A Study of Community Health Councils* (London: Centre for Studies in Social Policy, 1976).

41 The importance of consent materials in helping RECs assess applications is made by Emma Angell and colleagues who, in a study comparing the way in which three different RECs evaluated the same 18 applications, noted that, with 51 out of the 54 letters giving decisions about the applications raising such concerns, 'Issues relating to informed consent formed the ethical concern most often raised': Emma Angell, Clare Jackson, Richard Ashcroft, Alan Bryman, Kate Windridge, and Mary Dixon-Woods, 'Is "inconsistency" in research ethics committee decision-making really a problem? An empirical investigation and reflection', *Clinical Ethics*, 2 (2007): 93.

42 Laura Stark, 'Reading trust between the lines: "Housekeeping work" and inequality in human-subjects review', *Cambridge Quarterly of Healthcare Ethics*, 22 (2013), 394.

2

Trust, local knowledge, and distributed centralisation

'I remember when I went on a training course they did suggest that one of the ethical issues you were supposed to consider is the qualifications of the person doing the research. Now that's a hell of a lot easier if you all know who this person is.'

Celia, expert member, St Swithin's Local
Research Ethics Committee (LREC)

The potential links between LRECs and local research institutions come into sharp relief before a meeting of the St Swithin's committee. Members are milling around, and the Chair tells an anecdote from the night before. While at a formal drinks party for hospital staff, the Chair 'was talking to a researcher – I think he was drunk – who said that, after the Chief Executive, I was the most powerful person in the Trust!' by virtue of serving as Chair of the LREC associated with a research-intensive institution like this hospital. But such influence is not just one-way; hand in hand with the Chair's perceived 'power' within the hospital sits the way in which St Swithin's shapes the experiences, procedures, and expectations of this REC.

In a contrasting example, Northmoor LREC is discussing an application that involves researching a condition prevalent in a specific ethnic group, which is well represented in the local area. As the lead reviewer, Amanda (expert) is largely happy with the application, noting that 'Northmoor is a good area to investigate this condition, but I do have slight concerns about the number of studies we do on these patients', a point raised by Frank, the Vice Chair who is running this meeting, with the applicant when he attends: 'Is there an issue of the over-research of these patients at this clinic?' This prompts a response from the applicant: 'I agree with you on the over-research issue and had the same thought, but we're uniquely positioned in that we have a [condition] clinic for these patients which has been running for a year and these symptoms are a major issue, and we need to sort out the symptomology.' While what is being discussed here – possible over-research of a specific community well represented in a specific area – is clearly about the 'local', it seems quite different to the institutionalised local concerns being discussed at the beginning of the St Swithin's REC meeting.

That local relationships and knowledge play a role in ethics review should not come as a surprise. The locally embedded nature of the research ethics review system in the UK is present in the titles of two of the committees I sat in on. Both St Swithin's and Northmoor were LRECs, *Local* RECs, raising questions about what it means for a regulatory body to be 'local', how 'local' concerns – whatever they might be – influence the kinds of trust decisions these groups reach, and how committee members see their role in the context of the local. Beyond this, there is a second kind of REC, a Multi-centre REC (MREC) like Coastal, set up to review research taking place at five or more different sites. These MRECs seem in some sense to *not* be local, or at the very least, to not have the same kinds of institutional links and knowledge of the local area as LRECs; how do decisions vary between the different types of committee?

This chapter seeks to unpack the idea of 'local' review which has longstanding roots in historical debates about RECs in the National Health Service (NHS), exploring how various versions of the local – as related to a specific institution, or as being about a specific geographic area, or as being defined in opposition to centralised control – have been mobilised and enacted in policy disputes and REC decision making, and in turn how local knowledge is drawn on in deciding on an applicant's trustworthiness.

In policy terms, the importance of geographic locality draws on the intuitive appeal of making decisions at the local level, emphasised by the passing of the UK Localism Act of 2011, which justifies changes to the powers of local authorities on the grounds that 'power should be exercised at the lowest practical level – close to the people who are affected by decisions, rather than distant from them'.[1] In the context of the NHS, however, attitudes towards localism (and its contrasting principle, centralisation) have historically been characterised by a degree of ambivalence. The centralising tendencies are clear – a national health service with national standards, treating everyone equally – well summed up by Aneurin Bevan's oft-quoted claim that if a bedpan drops in a hospital corridor, the reverberations should echo in Whitehall. At the same time, over the years, policy changes – for example, the introduction of general management or the 'internal market' – ostensibly decentralised power, allowing local organisations to respond to consumer needs in a more flexible manner.[2]

To a large extent, the localised (or not) nature of ethics review has taken place at one remove from these broader debates within the NHS. The structures and policy debates surrounding RECs have been flexible with regard to the value of local decisions about research – supporting localism in the 1980s and resisting it in the 1990s – often in tune with trends within the broader NHS. However, the role of what we might call (acknowledging the complexity and ambiguity in the term) 'local knowledge' within the decision-making processes seems consistent, whether or not the Department of Health (DH) approves of RECs exercising local autonomy.

In line with the other aspects of REC decision making explored in this book, the local character of these committees, at least in terms of institutional affiliation, is rooted in the earliest days of the REC system, and can be seen at the core of several key historical developments and debates. To underline the deep-seated role of local review, this chapter begins by exploring two key debates in the development of RECs: attempts in the mid-1980s to set up a 'national REC', and the arrival, in 1997, of MRECs.

As already noted, LRECs' origins lie in the hospital-based committees that arose from the Royal College of Physician's (RCP's) 1967 call for the setting-up of bodies to review medical research prior to it taking place. This was such a success that, by 1972, the government of the day could confidently claim that at least 238 such committees had been set up in hospitals around the UK, including all teaching hospitals.[3] The role of the relevant government department – the Ministry of Health before 1968, the Department of Health and Social Security (DHSS) afterwards – in the oversight and guidance of RECs was, at this time, extremely light-touch. However, what guidance they did offer tended to strengthen the local character of these bodies, at least in terms of geographic localism; in its 1975 endorsement of the RCP's 1973 follow-up report, the Department went back on its previous position regarding lay members, accepting their role on RECs, with the 'Secretary of State [asking] Area Health Authorities to consider choosing a member of the appropriate Community Health Council, as a lay member'.[4]

Community Health Councils (CHCs) were created to fill a gap caused by the 1974 reorganisation of the NHS. In altering the makeup of hospital boards, these changes removed an important strand of lay management, resulting in 'a conspicuous absence of anything remotely resembling consumer representation' and highlighting a political conundrum: 'how to reconcile the emphasis on centralised planning with the currently fashionable rhetoric of local participation'. The CHCs were created to resolve this issue, to both 'supervise the running of services' and act 'as one of the channels through which local people ... [could] ... keep the area health authority informed of any problems'.[5] Composed of people nominated by local authorities, voluntary organisations, and Regional Health Authorities, CHCs were a good source of lay members for RECs. They provided articulate, 'committee-savvy' people who had an obvious interest in the NHS. These were also people with close ties to local healthcare and a role that, at least in part, centred on representing the needs of local healthcare users. As Barbara Castle, the Secretary of State for Social Services, noted in 1974, CHCs were forums 'where local participation in the running of the NHS can become a reality'.[6] Thus, through this guidance, and perhaps unintentionally, the DH ensured that RECs drew their professional membership from the individual hospitals in which they were based, while their lay members were chosen from bodies with ostensibly strong geographically local ties. While committees themselves were set up in the context of specific institutions, these changes introduced an important set of members (CHC-nominated lay members) who were not institutionally located but rather represented the interests of healthcare users from the local area.

While these committees were in line with the Department's vision of ethics review remaining (largely) the responsibility of the medical profession and beyond the remit of politicians or civil servants, the downside of such a large number of independent regulatory bodies started to be seen. The medical press began publishing papers criticising ethics committees on a range of different issues, including the effects of having to apply to a wide range of different committees and the variation in response.[7] This in turn raises the third kind of localism present in these policy discussions, defined in contrast to centralised ethical review.

As the organisation that had ended up responsible for the development of RECs in the UK, the RCP saw its role as one of coordinating these disparate and diffuse bodies in an attempt to ensure consistency and avoid disrupting medical research. This was mainly done through regular meetings of Chairs of NHS RECs, which began in 1974 in the wake of tension over the RCP's 1973 report (which recommended against setting up a central committee to oversee the activities of RECs) and the views of a number of senior members of the College who were in favour of a central REC. The solution was to periodically bring together the Chairs of RECs to debate issues of practice such as the composition of committees, the regularity of their meetings, and the use of chairman's action.[8]

Yet by the mid-1980s, other organisations, specifically the British Medical Association (BMA – the trade union and professional association for doctors in the UK) had also moved into this area of debate and had begun to push for a quite different, far more centralised system of ethics review. The Central Ethical Committee (CEC) of the BMA began to debate these issues, as well as broader topics related to RECs. Following a discussion paper presented to the CEC in March 1980, the BMA announced in March 1981 that its governing Council had approved a report making recommendations about both the remit of such committees and their membership.[9] Two years later, 'no uniform action seems to have followed' the 1981 report, and the CEC re-entered the debate. In addition to reiterating the need for a more coherent system of local RECs, the CEC, echoing debates rejected by both the MRC and RCP from a few years earlier, now argued that 'there is a need for a national (UK) committee for ethical research' to review clinical studies taking place 'in more than one area'. Such centralised review of multi-centre research 'would avoid the danger of the same protocol being approved by one Local Ethical Research Committee (LERC) and rejected by another'.[10]

Emboldened by the BMA Council's 1984 approval of this revised report, over the next 12 months, members of the CEC met representatives of a number of professional groups – for example, the

Royal College of Nursing, Royal College of Midwives, the British Psychological Society, and British Association of Social Workers – encountering broad support for the idea of a national REC to review multi-centre research while at the same time encountering concerns about medical dominance and the funding of such a committee.[11] In particular, representatives of CHCs expressed concern about the potential hierarchical nature of the relationships between local RECs and a central committee, and the possibility 'that Local Ethical Committees would be effectively down-graded if a National Committee was set up to look up "spectacular research" '.[12]

Despite the support of these professional bodies, the BMA was having trouble generating much interest on the part of the DHSS; in a letter to Sir James Gowans, Secretary of the MRC, J.D.J. Harvard of the CEC complained that 'the BMA has been trying, without a great deal of success, to get the DHSS to become actively involved in promoting an effective network' before proposing a meeting with the MRC 'to discuss how we might make progress on both the setting up of an effective network of Local Ethical Research Committees and on the proposed National Ethical Research Committee'.[13] This lack of interest on the part of government is unsurprising. In 1982, a new wave of NHS re-organisation moved on from the 'values of efficiency and rationality' embodied in the 1974 changes, seeking rather to 'stress ... the virtues of localism and small size'. As health minister Sir George Younger told Parliament: 'The thrust of our policy ... is to have decisions taken as near to the point of delivery of services as possible.'[14] The BMA's push for a centralised REC came at a time when the overall NHS was moving away from centralised control and towards local responsibility. And it is not only with the DHSS that their case was swimming against the tide: scepticism on the part of the MRC can be found in the handwritten notes made on the copy of the CEC's 1984 revised report included with Harvard's letter to Gowans. To the claim that 'An adequate system of local ethical research committees (LERCs) is required' the note asks 'Is the present system inadequate?' and the claim that 'Finally there is a case for a National (UK) Committee' is dismissed with: 'What is the case?'[15]

Over the course of 1985, the BMA's attempt to reform RECs and set up a national committee to review multi-centre research came undone. Its ill-timing with regard to broader NHS changes around de-centralisation came at the same time as the RCP – the professional body that to this date had taken responsibility for RECs – published the first of its periodic guidelines for ethics review, something that James Gowans pointed out in his reply to J.D.J. Harvard, and which an internal MRC note meant was 'doubtful whether any BMA initiative was required'.[16] Certainly, in a discussion with the MRC, Raymond Hoffenburg (President of the RCP) made it clear that he would like to see the 'results of the new initiatives [i.e. the RCP Guidelines] before any changes are discussed' and that he was 'generally unhappy about the BMA involvement'.[17]

In a meeting with representatives from the BMA's CEC, the MRC's case against the national committee 'emphasised the need for ethical matters to be decided amongst colleagues at a local level and questioned the type of national committee which had been proposed'. The MRC's hand was strengthened when 'it became apparent during the discussion that they [i.e. the BMA] knew nothing about systems in Europe or of the activities undertaken by the Royal College of Physicians'.[18] The MRC's appeal to the primacy of local ethics review – in this context defined in opposition to a centralised committee – is made even clearer the following month in a reply to a letter from a Mr Cropp, Chair of the Ethical Committee at East Hertfordshire Health Authority. In his letter, Cropp asks whether the MRC had considered setting up a national ethical committee to review multi-centre research, largely on the grounds that this would save time and money at the various local committees. Malcolm Godfrey's reply is explicit in its defence of local review:

the MRC and the research workers we support might find it convenient to have a national body to approve national studies; but the convenience would be outweighed by the strength the present system derives from the immediacy and autonomy of local committees.[19]

What is meant by 'local' here is less to do with the specifics of different parts of the country, and more to do with the institutions (hospitals mainly) within which research took place. The knowledge and oversight of fellow doctors working at the same hospital are presented as a crucial part of the regulation of medical research. The paradox is that, as shown later in this chapter, the 'local autonomy' of RECs, so vital in the mid-1980s to defend medical research, became, in less than a decade, a dangerous impediment to scientific progress in need of restriction and reining in.

In addition to these external challenges to the centralisation of ethics review, the CEC found its position undermined from elsewhere within the BMA. In November 1985, a sub-committee of the BMA's own Central Committee for Hospital Medical Services (CCHMS) 'took the view that there was no need for a National Ethical Research Committee'.[20] When the CEC ignored this dissent, produced another paper (its third report on the ethics review system), and got it approved by BMA Council, a second sub-committee of the CCHMS raised objections, and the CCHMS itself 'asked for the CEC's paper to be redebated by Council', as did the Medical Academic Staff Committee (MASC) which 'had objections in particular to the concept of a national ethical research committee'.[21] In March 1986, the BMA Council agreed to reopen debate – essentially withdrawing its approval for the CEC's report – asking that the 'motion approving the committees should "lie on the table" so that further consultations could take place'. The concerns of both the CCHMS and the MASC were voiced by a J.P. Payne who 'objected to a national committee in principle. The representation was unsatisfactory and who would it be responsible to? The important thing was to get the representation on the local committees correct.'[22] These concerns from other parts of the BMA were rooted in both the interests of researchers and the proposed importance of local review, with one critic noting 'the association was going down a road which did not have the support of many people actually involved in research', and another pointing out that it was hard to see 'an effective role for a national committee that did not disenfranchise the local group'.[23]

When in September 1986 the CEC met with representatives of the Royal Colleges, objections centred on the now familiar problems of undermining local RECs, although the emphasis here was on the role of geographic locality, with such committees being 'important for assuring the local community that research was being carried out in the correct way and with regards for the local area and its particular considerations'. The problem with a national committee was that it 'might become too powerful' and that a 'local committee might feel pressurized by a national body ... when the local committee may indeed have the best knowledge to decide whether a protocol were appropriate for that area or not'. Despite the best efforts of the supporters of the CEC position, the representatives of the Royal Colleges remained suspicious, concerned that 'a national ethical research committee would become a "big brother" type body'.[24] While over the next few years the CEC continued in its attempts to enlist the support of other bodies for the development of a central REC, it still found the Royal Colleges 'apprehensive' and intransigent. In their 1990 revised guidelines for RECs the RCP ignored these pleas for centralisation.[25]

These debates suggest that a key aspect of resistance to the BMA proposals lay in the rivalries and tensions between the bodies responsible, in various different ways, for the oversight of clinical research and the behaviour of medical professionals. In the case of the MRC, while it is clear that they accepted the need for improvements to the REC system, they were reluctant to allow the BMA (essentially a political, lobbying organisation on the part of doctors as a whole) to shape these changes. What is interesting is the way that opponents of the BMA's suggested changes draw again and again on the value of local ethics review, the idea that somehow local committees – by virtue of their knowledge of the local setting and people (or perhaps of doctors within a specific hospital) – make better decisions than those that might be made by a putative central, national REC. That the motivations of much of this resistance lay in an institutional 'turf war' does not undermine the persuasive power of appealing to local autonomy.

While the possibility of a single central REC had been avoided, it had become clear that the DHSS could no longer duck its responsibility for research ethics review in the NHS. Taking charge, the Department published the 'Red Book' in 1991, a guidance document formally locating RECs within the NHS, officially designating them as LRECs, as well as offering up guidance for the membership of such committees. These guidelines included an important element of de-institutionalisation, suggesting that, rather than specific hospitals, it is district health authorities – mid-level bodies that covered populations of between 200,000 and 500,000 people – that should take responsibility for establishing LRECs and providing them with administrative support. In this way, the Red Book attempted to define 'local' in exclusively geographical terms.[26]

Despite its standardising intentions, the Red Book says little about multi-centre research. It simply suggests that, while each individual LREC is free to review research taking place at more than one site:

> It would, however, obviously be sensible – in the interests of eliminating unnecessary delay and of ensuring that similar criteria are used to consider a proposal – that committees should arrive at a voluntary arrangement under which one LREC is nominated to consider the issue on behalf of them all.[27]

In addition to this example of wishful thinking, almost as soon as the Red Book was published, researchers from Swansea University were asked to report on the issues around the ethical review of multi-centre research, a project that produced a report and set of training materials for LREC members in an attempt to produce a degree of standardisation.[28]

Yet, perhaps unsurprisingly, complaints about variations in approval for the same research projects continued, to the extent that even a civil servant could state that:

> from the point of view of those people who sponsor multicentre trials ... [having to apply to multiple RECs] makes their life very difficult and it is understandable why they become concerned that something should be done to improve the current system as far as multicentre trials are concerned. This was an issue raised to the DoH [Department of Health] by the pharmaceutical industry, the Medical Research Council, medical research charities and also the DoH itself.[29]

Perhaps the highest-profile case for revision of the ethics system was made by the diabetologist and later President of the RCP, George Alberti, who, in an editorial for the *British Medical Journal*, highlighted the 'large variations in practice' between committees and the resulting 'groundswell of complaint from research workers, frustrated at the delays incurred, the cost involved and the unnecessary duplication of effort'. Suggesting that 'a consensus has emerged' around the need to coordinate committees, Alberti argued that 'committees should retain the ability to turn down trials on local grounds, although the rugged individuality of some committees requires tempering'.[30]

This tension between the rights of individual local committees and the need to restrict (in some way) their decisions is reflected in the solution reached by the Department: to set up eight MRECs which would review research taking place at five or more different sites. According to one of my interviewees who worked in industry at the time, the key goal of this new system was that 'if somebody wanted to do a project on multi-site, it didn't want to go to 121 committees' – a goal 'largely driven by the fact that the Minister sees pharmaceutical research as a good money-spinner for the UK, and so they've tried to tighten up on the time limits, and to try, to slacken down on the bureaucracy'.[31] The regionally named MRECs – South East MREC, London MREC, North West MREC, and so on – swiftly ran into problems. The pharmaceutical industry, the putative beneficiary of the MREC system, was uncertain about the new committees and expressed a preference for those LRECs they had come to develop a productive working relationship with. As a result, according to this same interviewee, firms tailored their research to their preferred regulators, going 'out of their way to keep to four centres or less [for their research] so that they wouldn't have to go to an MREC'.

A far more serious problem lay in the tensions that developed between the new MRECs and the pre-existing LRECs, tensions that some prescient observers had predicted prior to the MRECs' arrival.[32] As one interviewee from Coastal MREC put it, 'there was an awful lot of antipathy toward the MREC system ... because obviously there was this sense of being controlled from the centre, that it was being driven by big pharma ... that really just wanted us to rubber stamp everything. Local autonomy would be lost'. The problem lay in the process mediating between the two kinds of committee. An applicant considering multi-centre research – i.e. that 'carried out within five or more LRECs' geographical boundaries' – was expected to apply to the MREC in the region in which the principal researcher was based. If an MREC approved the application then its opinion was sent to all the LRECs in the areas where the research was due to be conducted. These local committees could then choose 'to accept or reject the protocol for local reasons but ... [were not] ... allowed to amend it'. The only exceptions to this 'no amendment' rule were '*purely locally applicable amendments* which do not affect the integrity of the protocol', the usual examples requiring 'information sheets and consent forms in minority languages' common in those local areas.[33]

The result of these peculiarly balanced powers – denying LRECs the right to formally review these applications but allowing them to reject or make limited amendments – was that:

> LRECS having been deprived of the right to review MREC-approved studies in full were in more or less open revolt, because they saw many MREC-approved studies as being full of ethical holes that the LREC could identify, but could only do something about by guer- rilla tactics of delay or non-approval on spurious 'local' grounds.[34]

The committee that recommended setting up MRECs – the Chief Medical Officer's Consultative Group on Research Ethics – was aware of the tension around the need to maintain the autonomy of LRECs. This body was brought together in late 1994 to draw up plans for the review of multi-centre research, starting from the position that local knowledge mattered in ethics review; one of this group's early discussion documents noted that there was 'almost unanimous agreement that under whatever new system is eventually adopted, it will be necessary for multi-centre trials to be referred to every LREC involved to consider local implications'. While there was some support for a central committee model, locally based and regional alternatives were also proposed. In discussing these options, the document set out the need 'to consider how the confidence of LRECs in such a system could be developed to minimise "maverick" local refusals', noting the need to 'reduce the incidence of LRECs re- jecting trials purely because the advice/lead was seemingly imposed from on high'.[35] In the broader policy context, while this group pre- dated the New Labour government of 1997, the centralising ten- dencies of its eventual recommendations were fully in keeping with the new administration's approach to the NHS.[36] Although the need to preserve the rights of LRECs to exercise local judgement over MREC-approved applications is present throughout the documents discussed by the Consultative Group, it is clear that not all members of the group were convinced. In a letter to the civil servant coordin- ating the group, one member complained that:

> The discussion this morning appeared to confirm that the dom- inant assumption among members of the consultative group is that turbulent LRECs can only be constrained from above, such as by

McRECs [i.e. Multi-centre RECs] which are actually more expert and powerful but do not wish to claim or appear to be so.

The letter goes on to suggest that 'A more cost effective and psychologically astute approach would be to start by trying to trust LRECs. They will continue to be very influential in any future system.'[37]

Despite this awareness, the early days of the MREC system were characterised by exactly the kind of resistance from LRECs the Consultative Group was afraid of. Echoing previous changes to the ethics review system, medical journals began to publish papers highlighting the variation between LRECs in their responses to MREC decisions, and the cost of copying and posting numerous copies of applications and MREC approvals.[38] The actual number of outright rejections of applications was low – one survey suggested only 1.5% of MREC-approved applications were actually rejected by LRECs, usually on the grounds of unsatisfactory information sheets. Researchers tended to be more concerned with delays in approval and amendments.[39]

A more sympathetic perspective came from some of my interviewees, with one LREC member pointing out the advantages of some local review of multi-centre research:

> you knew your local patch and there was a time when we thought, 'gosh, this GP must be giving every woman over 45 hormone replacement treatment therapy in some trial or other', and we began to question whether he was actually being totally honest with the subjects

while at the same time acknowledging the benefits of the formalised system that came with the MRECs:

> The MRECs had to have certain formal members ... like statisticians, to make sure that all those [applications] were dealt with by someone who was properly qualified, instead of having a haematologist who happened to know a bit about statistics, doing the statistics.

Similarly, a member of Coastal MREC noted that local review brought certain insights an MREC could not, accepting that:

> local people might say, 'well, hang on a second, this guy is now doing ten projects ... or he's doing projects on the same population

> ... or he's a total cowboy, we wouldn't trust him to cross the road
> ...' or whatever it is and they would be able to have that kind of local
> perspective that we would not have.

Over time, initiatives such as joint training and meetings between
chairs of RECs did much to reduce the tensions between LRECs
and MRECs, but perhaps the main impact came from structural
changes within the system. I began my fieldwork just as LRECs
were undergoing a re-categorisation, with some committees remaining focused on research within their local area (Type 2 committees) and others being 'upgraded' to the status (if not the title)
of MRECs, being allowed to review research conducted at any site
in the UK. Over time these Type 3 committees have become the majority within the NHS; of the 86 RECs now active in the UK, 54 are
classed as Type 3.[40] As already noted, over time ethics review has
become incorporated into international pharmaceutical regulation,
and the original goal of these changes was to implement the EU's
2001 Clinical Trials Directive, enacted in 2004, which, in the case
of drug trials, required a single review in each EU member state
where the trial was being conducted. While some member states
opted to set up single central RECs to review pharmaceutical trials,
the UK, in keeping with the broader direction of DH policy, chose
to empower a wide range of RECs to provide this single review.[41]

What these committees are *called* also, apparently, matters.
Over time, the 'L' and the 'M' have disappeared; the NHS now
only has 'RECs', further erasing the differences between different
committees. Going further than this, since finishing my fieldwork,
RECs have had their names changed to remove visible associations
with specific institutions. Initially names were changed to bland
area-specific titles, before being swapped back to something with
more geographically local character (though no institutional link).
Though true in a number of areas (Liverpool, for example), this is
most obvious in the case of the London teaching hospitals – 'King's
College Hospital REC' became the rather characterless 'South East
London REC 3' before being renamed as 'London – Dulwich';
'Charing Cross Hospital REC' became 'West London REC 2' and
is now 'London – Fulham'. Thus, over the past decade, the NHS's
research ethics review system has been characterised by a formal
'de-insitutionalisation' of review, a putative separation of ethics review from researchers and the institutions where research is carried

out, while trying to maintain some link to the local geographic area within which the REC is based.

While early debates made use of localism as a tool to deflect the BMA and reinforce the legitimacy of the RCP oversight of RECs, the formalisation of ethics review in the shape of the 'Red Book' solidified the interests of the pharmaceutical industry. It was no longer enough to protect the interests of local (institutional) researchers through local review; the economic value of industry research required some form of centralisation. Committees that started as creations of specific hospitals moved to the remit of district health authorities, found their role trumped by MRECs, became answerable to Strategic Health Authorities (with their much larger populations), and finally became fully national committees, able to review research taking place at any site(s) in the UK. Rather than this leading to a concentration of control at the centre – in the form of the long-argued-for national REC, for example – the solution adopted by the DH was to empower each individual committee. If a national REC is a committee that can review research for anywhere in the UK, then the changes introduced in the wake of the Clinical Trials Directive created over 50 such committees.

From the point of view of researchers and the DH this approach to organising research ethics review, a form of *distributed centralisation*, has a number of advantages. Such a system both distributes power (to individual RECs) and centralises it (away from the monopolistic rights of local committees to approve research taking place in their patch). Overlap between the remit of different RECs provides surplus 'capacity' within the system, making it more likely that in any one month all applications made to RECs will be reviewed promptly. Only requiring a single review for multisite applications (however many locations the research will be run in) reduces researchers' concerns both about variability of response between RECs and with regard to bureaucratic overload. However such changes also served the additional goal of, at least in theory, breaking the link between local committees and local populations of patients and volunteers. As one policy maker put it, drawing on the idea of local as being about place, not institution: 'In England, is someone in Worcester that different from someone in Gloucester, and do you actually need an ethics committee in both places?' From this point of view, the ethical issues to do with a particular piece of research are unlikely to vary geographically, and thus the need for RECs to be locally placed is unconvincing. In contrast to this position is Sarah Dyer's view, rooted in an understanding of the practice of ethics review, that:

> LRECs make judgements about locality issues all the time ... These judgements, though, are not made separately but enmeshed within a network of connected concerns. For example, a committee may have concerns that someone is doing too much research or does not have the right expertise to conduct particular research.[42]

The paradox of distributed centralisation is that the flexibility built into the system – allowing researchers to choose where to apply for review – enables researchers to deliberately choose to apply to their local REC, *strengthening* local ties. This might be the committee located closest to them or associated with their research institution. In the US, concerns have been raised over the issue of 'IRB [Institutional Review Board] shopping', the practice of submitting protocols to multiple IRBs until one is found that will approve the protocol.[43] In contrast there has been little discussion about the UK equivalent, the cherry-picking of specific RECs that became a possibility following changes in the ethics review system. Previous changes produced similar behaviour: as noted above, following the introduction of MRECs, many pharmaceutical firms limited the number of sites involved in any one trial, to ensure they were reviewed by trusted LRECs, rather than the unknown quantity of the MRECs. Despite the formal delocalisation of the ethics review system, institutional and local bonds remain, and what one might label 'local knowledge' (covering both geographical and institutional sense of local) continues to play an important role in trust decisions around ethics review.[44] The remainder of this chapter explores the role of such knowledge in REC trust decisions, and shows how it can underpin both routine decisions and pathological events such as the TGN1412 trial.

The role of institutional local knowledge in REC trust decisions is perhaps most obvious in one project reviewed by St Swithin's LREC, a multisite follow-up to a previous project run at the hospital. The research involves an innovative piece of monitoring equipment developed for use during a heart attack, the underlining basis of which is controversial. Not only is this second study to be run at a number of different sites but the equipment will be used in a slightly different way than previously, which may potentially make it more intrusive.

As the committee discusses the application, there is general concern among members in the shift between the first study, where there was no possibility of interference in attempts to resuscitate a patient, and the proposed new research where the researchers would be in the same room, potentially interfering in the resuscitation. One expert member, Craig, suggests that cardiac arrests are most likely to happen in A&E (i.e. the Emergency room) or the cardiac unit and in the A&E there *would* be a safety issue, since the resuscitation team might not be aware of the proposed research, whereas in the cardiac care unit, the equipment could be set up in advance and staff pre-warned. The Chair sums up the REC's position: 'I think we're coming down on the idea that unless he can make a strong case, I think we would only agree on research happening in a special unit.'

The applicant is invited into the room. While he is not the lead researcher on this study, he is the designated 'local researcher', an experienced, senior nurse who trains others in the hospital in resuscitation techniques. During the discussion with the committee, the researcher points out that limiting the project to heart attacks in the cardiac unit would cause problems for the research since, because of the prophylactic treatments used there, cardiac arrests are rare. The applicant states strongly that he would never allow the research project to interfere with his treatment of patients: very often he is working with a colleague, who would be treating the patient and that would free him up to set up the monitoring equipment being tested. Louise, an expert member, responds: 'I trust your professionalism, but I am trying to understand the process: if we were to say that we would only approve the study if two of you attend a cardiac arrest, would that be OK?' She then follows up with a number of questions, mainly exploring the attitude of other professionals involved in resuscitation to being included in the research. Following a brief demonstration of the equipment to be tested, to show its non-obtrusiveness, the applicant leaves the room, and the committee continues to debate the study.

Celia, a GP member (and hence, non-institutional, unfamiliar with hospital staff) is still not happy about the chance that setting up the equipment will interfere with clinical practice. In response, Craig notes: 'I know him [from the intensive care unit] and have no concerns about his priorities regarding patient care, but as a committee we should not have to rely upon *his* professionalism. The study design should remove our need to know the individual involved.' More importantly, since this study is to be carried out at a number of sites across the country, 'we have to ignore our knowledge of the applicant' since the study would be, as a lay member puts it, 'run here by someone we trust and then expanded to the country as a whole'. The study would be largely unproblematic if it were *only* taking place in St Swithin's Hospital, since members of the REC know and trust the local researcher; but the multi-centre nature of the study means that it would also be carried out at a number of different sites. The study is rejected.

The key problem in this case was that the REC has no knowledge of the individuals involved at the other sites, and cannot attest to their trustworthiness. As Craig, the expert member quoted above, said later in interview, 'that became a show stopper, didn't it? Because so much of what was going to happen was dependent upon a particular personality.' While in theory the study design should have removed the need to rely on such situated knowledge, in practice, given the committee's concern about the possible risks to patients, knowledge of this applicant and his trust-warranting properties (his 'professionalism') and parallel *lack* of knowledge of the other people carrying out the research were the deciding factors.

These reasons become even clearer a few meetings later when the project is resubmitted to the REC for approval. The Chair (who is acting as lead reviewer) reminds the committee of a number of previous concerns – 'that the resuscitating officer spends time putting up apparatus rather than working on the patient. There are also consent issues...' – and finds the applicant's 'written answers unconvincing. Their argument is that the resuscitation officer is totally supernumerary to the [cardiac] arrest. They are there to help out to advise those actually doing the resuscitation. Are people convinced?'

There is a resounding 'No' from around the table, and members begin to contribute their views. Linda (expert member) suggests that 'we cannot be sure that the same procedures will be done elsewhere'. When challenged by another expert, Ingrid – 'That was a very categorical no. Surely the resuss. [resuscitation] officers elsewhere will put the patients' interests first?' – Linda underlines the importance of direct knowledge of the local researcher: 'We don't know, do we?' The role of such knowledge in risk–benefit decisions is highlighted by Louise (expert member) pointing out that 'We can only support this if it's absolutely benign to patients'; the benign nature of this equipment is dependent upon the qualities of the researchers using it. Without knowing who will run the research at other institutions, the committee cannot know for sure that the risk–benefit trade-off is justified.

Of course, compared to many other RECs, members of the St Swithin's committee have privileged access to knowledge about the researchers who apply to them due to the institutionalised nature of their committee and the close ties it has with the hospital that serves as its not-quite host. There is a perception that, somehow, the REC belongs to the hospital. As one member of the committee put it to me, 'The committee sees itself ... It isn't necessarily *of* those institutions [i.e. the hospital and related medical school/university], but very much *tied into* those institutions ... maybe pragmatically yes, that's where the bulk of our stuff comes from.' Malcolm, the Chair of Coastal MREC, making a distinction between LRECs and his own committee (which, of course, has never had institutional links) suggests:

> LRECs are, to a greater or lesser extent, affiliated to their local hospital trust. They sit in trust premises (in the main) and they look at a lot of research from that specific trust. And, of course, historically, they developed as ethics committees for the trust or the district health authority or before that, the district hospital. They were essentially owned by the local hospital, in the same way as IRBs are.

Members of St Swithin's REC also perceive the hospital's attitude towards the committee in proprietorial terms: 'that St Swithin's Hospital feels that it has ownership of the REC, to some extent, which is probably not accurate but I think they see it as part of the organisation'. In terms of its formal status, the REC is, of course, independent from the hospital, something that sometimes makes its way into committee meetings. For example, on one occasion, an expert member complains about the way recruiting posters for studies such as the one under review were stuck up in lifts, partly because they tended to remain up past the end date of the study but

also because they looked 'tacky' – as another member put it, 'A bit
like the telephone boxes in Soho!'[45] The committee administrator
points out that it is not the REC's role to give guidance on this sort
of thing; it was an issue of hospital policy and 'We need to signal
that we are not a hospital committee.'

To some extent this perceived 'ownership' of the REC is related
to the high proportion of hospital and university staff who sit on
the committee, but this is about more than just numbers.[46] As the
vignette at the beginning of this chapter suggests – where the Chair
of the committee is described by a colleague as 'the second most
powerful person in the hospital' – serving on the REC is seen as
serving the hospital. In interview, Howard, the Chair of St Swithin's
LREC, reiterated the professional value of his role:

> There is also a sense towards which I am definitely, to some extent,
> loved by my institution because I'm Chairman of the Research Ethics
> Committee. It is a position that is seen as being an important pos-
> ition within [associated University] and within St Swithin's Hospital.

Such value is not just ascribed to chairing the REC but also to
simply being a member. The time members take out of their professional duties is accepted by colleagues because REC meetings
are seen as serving the interests of the broader institution: 'being
a big research institution, if I was heading off once a month on a
Monday afternoon to [another hospital] or another institution to
their Ethics Committee, I don't think it would be so favourably
received by my colleagues' (Linda, expert). The recruiting value of
institutional links is acknowledged by Howard, who noted that:

> some members of our committee you would not get serving on a
> generalised ethics committee ... they want to serve their institution.
> They can either sit on the ethics committee or they can sit on the
> Clinical Governance Committee or they can sit on some other com-
> mittee, and yet if they sat on the committee that was in charge of,
> you know, in charge of nothing to do with the hospital, they would
> lose their kudos, why should they do it?

Even the origin of applications reviewed by this committee is seen as
valuable at an individual level, in terms of the institutional intelligence
such information gives members: 'would I be on an Ethics Committee
if it wasn't local research? Almost certainly not because I wouldn't get

the same sort of interest and understanding in what's happening in the organisation, so ..., so it's not that altruistic' (Ingrid, expert).

Far from being unusual in the closeness of these institutional links, Howard suggests that St Swithin's REC is typical of a *specific subset* of committees, being 'very similar to the relationship between the other major teaching research intensive hospitals ... they tend to have very, very close relationships with their hospitals, because they are like us and have sprung out of an *institution*'s research ethics committee'. Thus RECs with strong institutional connections clearly generate local knowledge, dependent in part upon the network of relationships the committee members have with researchers working at the same institution. This knowledge is partly about people's research and the facilities available to support applicants (resuscitation facilities where there is a risk of anaphylactic shock, for example) yet it is also knowledge about applicants themselves, about their characters and qualifications.

However, as noted earlier in this chapter, in addition to being about institutional connections, in the context of ethics review, local can also mean 'geographically close', with ties to the same area as a particular REC.

The best example of the benefits, and tensions, produced by close geographic relationships comes again from St Swithin's, and the work of Mark Jones, an academic based not at the hospital but rather at a nearby university, with a very active programme of research and who applies every couple of months to the REC for new approvals. I first became aware of his work early in my fieldwork, when the committee turned to a new application, described by the Chair as 'complex'. While Jones' work normally comes to the St Swithin's committee, in this case the application was originally submitted to another REC who, while they accepted most of the study, were concerned by the underlying method which involved injecting complex organic molecules into healthy volunteers without prior approval from the Medicines and Healthcare products Regulatory Agency (MHRA), the drugs regulator (equivalent to the US Food and Drug Administration). Because this research focuses on basic physiology rather than drug development, there is no requirement for MHRA approval, but the initial REC was concerned about possible toxicology, and rejected the application.

Part of the problem, Howard feels, is that the ethics application form does not have enough space on it to discuss the safety issues involved in this research, and the other REC lacks the St Swithin's committee's experience in dealing with this principal investigator's research. Thus this application comes to the St Swithin's REC as a sort of appeal: it has been revised in the light of the comments from the other committee (i.e. it contains more toxicological data, therefore it is technically *not* an appeal), but is substantially the same application.

Following this introduction, Howard leads on the application itself. The overall design, number of visits required of volunteers, and the reimbursement for their time, the Chair notes, are 'along the lines of many studies we've seen from him before', but there are inconsistencies between the participant information sheet (PIS) and application form, and there is a need to ask about the exact dose of these biological molecules being used. At this point, one of the expert members – Ingrid – points out the unusual nature of such an appeal: 'It's very odd if we say that if one REC doesn't like it we re-send it to another.' Howard concurs: 'I agree, I feel a bit uncomfortable, but I think we should take this as an opportunity to look at the toxicology issues', essentially asking members to review the way the REC has thought about the toxicology of this principal investigator's research overall. In response, Ingrid notes that 'in the beginning we had similar concerns about the MHRA and toxicology, but we just got "worn down" by the sheer number of applications'.

Discussion moves on to toxicology, with the Chair clearly not happy with the other REC's 'expert opinion', although as Abigail, a lay member, points out, this study does raise novel issues since this project uses a synthetic compound rather than the naturally occurring one used in previous studies from this researcher. There are issues to raise with the applicant but clearly the committee does not have the same safety concerns about this research as the first REC that reviewed it; St Swithin's know from previous experience that this researcher has done this kind of study many times before, and that there is a great deal of supportive toxicological data underpinning this study's safety, much more than could be fitted into any one application. As the applicant is asked to attend, Ingrid suggests that, since 'we're alerted to the fact that someone else doesn't like it we might feel a little guilty about our approval of this'.

The applicant attends, and the Chair begins by pointing out a number of inconsistencies between the application form and the PIS. When discussing the safety grounds for stopping the trial, the applicant draws on his team's long history of work in this area: 'Our experience is, and we're done this an awful lot of times now ...'. In line with other occasions when this applicant attends, he is non-confrontational, he nods along with Howard's comments, and while he is emphatic about the safety of the compound being tested, he does not seem to regard these questions as illegitimate. He states he is happy to add additional immunological measurements to counter the chance of a reaction to the compound, although personally he is not convinced of any problem.

Following his departure, Ingrid smiles: 'He's a very smooth operator, isn't he?'

Linda: 'It's useful for us that this has been rejected by another committee, to reassess our decisions.'
Ingrid: 'To show we are not a push over!'
Linda: 'He wants to keep us onside.'

Celia then asks Ben, the committee's administrator, about the process that led the original application going to the first REC: 'How did it go to [other REC]? Did COREC [Central Office for Research Ethics Committees] decide that he had to apply outside his home turf?' Ben: 'Probably our slots were full' (this does turn out to be the explanation).[47] Celia's suggestion – that representatives of the DH in the form of COREC required Mark Jones to apply to a different committee – highlights members' awareness of their relationship to him, their knowledge of his overall programme of research, how that shapes their understanding of his work (and its risks), their willingness to trust him, and how this all might be seen from the outside.

This familiarity with Jones' work – this local knowledge – crops up in all assessments of his (frequent) applications, such as the one a couple of months later. Louise leads the review, starting by introducing the study in the light of everything the REC knows about his research, pointing out that the committee has approved multiple studies for two of the compounds in question, although none of these studies was a clinical trial; they were only looking at the physiological impact. The study under review gives both compounds together to see if they inhibit specific physiological processes. Louise reiterates that 'This is very similar to studies we have seen before', describing the similarities: 'It uses healthy volunteers, they get screened ... We've questioned him on this before and we've been satisfied with his response. They will do a pregnancy test. There is also a full-page reassurance on how the compounds are made.' When it comes to the PIS, Abigail is full of praise – 'He's got it down to a fine art' – as was Louise: 'All my comments in the margin were "good" and "cool".'

One mechanism for getting to know applicants is, as Sarah Dyer has noted, via members' experience of working alongside researchers:

> departments that are particularly research active are likely to be required to carry a proportionate responsibility. This dynamic ought to lead to local committees weighted toward the research specialism of that area. (Meaning also LRECs are more likely to know researchers more directly.)[48]

However, while there is clearly a relationship between Mark Jones and the St Swithin's LREC, in this case it is the result of frequent applications rather than being based in the same department. Although Hilary, an expert member, 'feel[s] I have a relationship with him coming here', in institutional terms she is 'not actually involved in any of their work ... [Despite this] ... They do, I feel, tend to take things on board, and it would be hard to say that there's ever been a time when you felt that they hadn't, that there was an issue.' However members are aware of how the way in which they treat these applications might be seen by the outside world; as we saw there is concern (probably misplaced) that COREC even reallocated this application first time round to an alternative REC. Members are also aware of the value of this kind of situated knowledge, about both the applicant and his research programme, that they have accrued over the time he has submitted his applications. With regard to Jones himself, Adrian (expert member) emphasises the value of knowing an applicant: 'It's just a natural thing, you say okay he's very trustworthy because he's doing good research [that] appears to be ethical, you ... have a bit more faith in him than somebody new who you don't know.' The circularity of this point, that Jones is trustworthy because of his 'good' research that is also 'ethical', highlights Steven Shapin's claims about the integrated nature of science and ethics, that prising apart the qualities of science from the moral nature of the scientist carrying it out is harder than it looks.[49] As Chapter 3 explores, much of Jones' trustworthiness flows from his willingness to attend the REC and the qualities of his performance once there, but the committee's view of his trustworthiness also draws on members' previous exposure to his research. More generally, Ingrid agrees that:

> the members of the LREC will of course know the people applying personally, and I'm quite sure that affects the way you think about it. If you know somebody to be a really good person with genuine concern and who's moral and ethical in themselves, then I don't know whether that influences ... If you know somebody's a bit of a cowboy, you look at it more carefully.

In characterising the nature of the knowledge of a researcher being described here, we might draw distinctions with the previous example, that of the novel heart monitor and the senior resuscitation nurse. In that case, the knowledge in question was explicitly institutional, derived from working alongside the researcher in clinical settings. In the case of Mark Jones, however, the source of the knowledge has little at all to do with REC members working alongside the applicant (none of them do) but rather repeated appearances before the committee – more a feature of geographic proximity. Thus, mapping in turn on to the ambiguity present in historical debates drawing on the concept of 'the local' in ethics review, RECs' local knowledge of applicants can be generated from different routes, from working alongside someone in a clinical situation or from their geographic proximity and hence opportunistic regular applications to the same committee.

In interview, Jones himself was reluctant to describe this situation in terms of a 'relationship' between himself and the committee, a concept he saw in negative terms – 'I'm friendly with the chairman or something and he's giving me special favours, I don't think that's true' – preferring to frame it as a case of 'expertise'. Yet St Swithin's LREC is not obviously more expert in terms of his specific research area than the first committee this application went to, where it was rejected. Neither committee has anyone qualified in this specific discipline. Any expertise the St Swithin's committee has gained is by virtue of this applicant repeatedly submitting research studies to them over a period of time, what we might call, drawing on the work of Harry Collins and Rob Evans, 'interactional expertise'. Typically, this is gained via socialisation with members of a specific expert group, and allows one to engage in informed discussion with such experts.[50] In this case, St Swithin's LREC displays expertise gained through the committee's relationship with this researcher, knowledge that results from regular long-term exposure to his idea and work, itself a result, less of working at the same institution and more about physical proximity and Jones being able to both apply and attend meetings of the same REC on a regular basis. This researcher's perception that St Swithin's LREC is somehow expert in his work is made explicit in a discussion with Howard, the committee's Chair, over the newly set up REC based in the university the hospital is affiliated with, and whether future applications should go there rather than the NHS REC. Jones' response is categorical: 'A non-expert committee does not make good points. This committee [i.e. St Swithin's] has built up a body of understanding and I would prefer that future studies come back to *this* committee.'[51]

As suggested above, members of the St Swithin's REC hold a range of nuanced views concerning the values and risks associated with such local knowledge and the kinds of relationships that develop with researchers based at local institutions. In comparative terms, as suggested by the Chair of Coastal MREC earlier in this chapter, this REC and those like it share many characteristics with US IRBs. However, while the institutional nature of IRBs has led to discussion of issues such as financial conflict of interest, and the way such bodies can serve to protect host institutions reputations' from embarrassing or controversial research, little attention has been paid to the way that IRBs are (externally empowered) internal self-regulatory mechanisms, where internal, collegial aspects provide the basis for trust decisions.[52]

The TGN1412 trial provides a good example of the way in which local knowledge can lead to pathological trust-based regulatory decisions. In Chapter 1 I argued that the Brent REC's previous experience with applications from large pharmaceutical companies shaped members' expectations about the safety of clinical trials as a whole. Yet committee members suggest their expectations were not just based on a generic understanding of industry competence, but also specific knowledge about, and experience with, Parexel, the company running the TGN1412 trial. As a member of the Brent REC put it:

> What people would have taken great comfort from I guess is that it [the trial] was being run through a clinical trial unit that people knew, people were happy with, visited, knew the staff well, knew that if we ever had any concerns, we would pick up a phone and they would be taken incredibly seriously and addressed and so on and I guess to an extent people felt there was some, well, 'this can't be a fly by night ... It's come through from Parexel and that's okay.'

This perspective is reflected in the view of one of the people involved in the trial who suggested that:

> we used to say that Brent was our home committee because we went to them two-thirds of the time. We'd go there first because it's local, so it was easy for myself or the other investigators to attend. We knew the administrator so we could phone her up and say, have you got any [application] slots? And we didn't have to go through the, whatever the system is called where you could end up being sent to Aberdeen [i.e. the central National Research Ethics Service system].

The position being mapped out here is very much like that of a US IRB, or an NHS REC such as St Swithin's: regular applications to the same committee allow relationships to develop between a REC and researchers, which in turn encourages trust decisions. Yet the difference is, of course, that unlike many US IRBs (where there is an obligation to apply to the host institution's own review body) or St Swithin's LREC (where the majority of applications come from researchers at the hospital or associated medical school), Parexel has no institutional link to this particular REC. Rather, the regular re-applications and subsequent development of a trust relationship between this committee and applicant are the results of broader structural changes.

One key aspect of the REC system that encourages the development of this trust is the paradoxical nature of the flexibility that allows researchers to apply to any one of a large number of different RECs. From talking to policy makers, it is clear that the aim of this flexibility was to prevent delays and allow applications to the REC with the soonest scheduled meeting. However, as already

Trust in the system

noted, researchers have also used the freedom to apply to *any* REC to regularly apply to the *same* REC (or small number of committees) in which they have confidence. While this is true of all researchers, changes in the organisation of pharmaceutical research have an additive effect to the impact of being able to choose the REC one applies to. Over the 1990s, the pharmaceutical industry was characterised by the rapid growth in the number of contract research organisations (CROs) which service the pharmaceutical and biotechnology industries by specialising in recruiting healthy volunteers and patients and setting up and running clinical trials. By outsourcing these activities to CROs, the companies that actually develop drugs – such as TeGenero, the firm that developed TGN1412 – save money and time. By the year 2000, CROs were running over 80% of Phase I trials in the UK. Whereas in the past, a number of different pharmaceutical companies might have applied directly to a REC for approval for their Phase I trials, now they were more likely to employ a CRO which, while acting on behalf of a number of different firms, serves as a 'funnel', making far more applications to any one REC than any individual pharmaceutical company.[53] The repeated applications from Paraexel to the Brent REC, and the subsequent development of trust on the part of committee members, should be seen as less of an isolated incident and more of an obvious, if unintended, consequence of changes within the ethics review system and the broader clinical trials industry. Since this close trust relationship arose out of broader structural factors, we should expect similar relationships to develop between other CROs and NHS RECs overseeing Phase I clinical trials.

While not all RECs have the kinds of institutional links or geographically proximate relationships discussed so far, in the case of those ethics committees which maintain (despite the various reorganisations of the system) a localised nature, or a close relationship with a specific institution, then local knowledge is an important way of helping the committee make trust decisions about specific researchers and their applications. If St Swithin's can be seen as representative of a specific sub-set of RECs (those with a long-standing affiliation to research-intensive medical schools), a key question is whether the kind of local knowledges crucial to *this* committee's trust decisions is at all available to other kinds of RECs, which may have a less exclusive relationship with a specific institution. In the case of Northmoor and District LREC, an expert member, Neil, noted that the change in the committee's status – from Type 2 to Type 3 LREC – allowing it to review applications from anywhere in the UK, was reflected in the localised nature (or not) of recent applications: 'I think the local side of it has begun to disappear as we're getting more studies trickling in which aren't from the local healthcare community.' However, he does point out that

> Having said that, still the majority are [local], I think that's one of the strengths of the system actually, because usually, when you take everyone on the committee, there's usually somebody who knows somebody who's involved with the research – either directly, indirectly, has worked with them in the past, whatever – and I think that's a beneficial link both with the committee and for the researcher.'

Since lay members often lack contact with researchers, expert members provide a crucial form of knowledge – that of other researchers – which is independent of their scientific or medical expertise.[54]

It is clear that Northmoor and district LREC also draws on institutionalised local knowledge to help shape its decisions, with this sometimes centring less on the qualities of a specific researcher and more on a detailed understanding of local clinical practices. In one case, the committee's review starts with Damien declaring a potential conflict of interest in the application – the researcher in question is a colleague of his. The committee is happy for Damien to remain involved.[55] The Chair, Martin, introduces the study which compares two different cardiac output monitoring devices. The aim is to compare the amount of blood pumped out, comparing the current gold standard, which involves inserting a catheter and is invasive, to a non-invasive new device. Because the study will take place in intensive care, the majority of patients involved in the study will be unconscious, raising issues around the PIS and consent.

The discussion is opened up to the committee as a whole, and Damien outlines his concerns: 'The problem with this study is that our unit *does not use* invasive monitoring that often, so the risk is that people will be tempted to use the catheter more often than normal' – it is clear that the REC is surprised by this – 'The unit uses a machine which is non-invasive but it's not calibrated/ gold standard.' Various suggestions are put forward to solve the problem raised by Damien: how to ensure enough patients are enrolled in this research without unnecessarily catheterising them. The obvious solution, suggested by one member, is to just recruit patients who definitely would have catheters fitted anyway. Martin highlights an alternative, asking: 'Are there other kinds of patients, who *could* consent, who normally get catheter monitoring?' Damien replies positively. The discussion centres on the PIS and consent issues. Megan then reiterates Martin's solution to the use of non-consenting intensive care patients: 'I would like it if you could do it on a group of patients who can consent beforehand, and who will be more likely to get the catheter.' The applicant attends and responds to the REC's questions but it becomes clear that the comparison of the two kinds of monitoring device is such that it is hard for the committee to see how the data will be adequately analysed, raising questions about the validity of the study. In the end, the committee rejects this application.

Obviously a range of factors influence the REC's decision in this case: a study involving patients who cannot consent in advance needs to have scrupulous information sheets and consent forms. As we will see in Chapter 4, a poorly designed study which cannot provide a statistically valid comparison between interventions will face considerable scepticism on the part of a REC. Yet underpinning all of this – highlighted by the Chair's suggestion that the hospital Research and Development office could monitor catheterisation rates to ensure there is no increase – lies the historically low levels of catheterisation in this unit and a trust-based concern about patients undergoing unnecessary interventions to make them eligible for the study. This concern comes directly from the local knowledge of one member of the committee, and derives from his working alongside the researchers involved.

Reflecting on such decisions, members of Northmoor LREC display the same kind of awareness of potential values and risks as those of St Swithin's. Amanda, an expert member, suggests that when familiar researchers attend the meeting (see Chapter 3): 'They're quite relaxed, they're not bound by the nerves bit, and they're kind of, everybody's familiar at least by sight, and there is a different feeling to it.' This familiarity is then followed by:

> there's often, 'oh well, usually their research is very good', so there's a kind of trust – there's a kind of trust relationship built up that you

might have a bit of tweaking to do but by and large you're fairly sure that this research is going to be both seen through to the end and have some kind of outcome to it that's beneficial to patients.

Of course, the reverse of this is also true: 'By the same token, there's a couple of researchers that come from the local hospital that people will say, "oh God, not him again", [laughter], or "this is always a bit dodgy", we always worry a bit about theirs.' At the same time, Martin, the expert Chair of the committee, notes that with repeat applications: 'people sometimes slip into comments like, "oh, well, we know that the counselling will be well done because it's this unit that's doing it", and it may be so, but actually, if you're not careful, you can slip into not scrutinising these because it's been done well in the past'. Thus there remains a tension – noted by members of St Swithin's LREC and at the heart of the TGN1412 case – between the insight into researchers' trustworthiness and abilities stemming from such knowledge, and the perceived dangers of too-close relationships between committees and researchers.

If local knowledge – either of an institutional kind or derived from familiarity arising out of geographical proximity – is important in LRECs of various kinds, we might expect a different story in the case of Coastal. As an MREC, this committee has never had any institutional affiliation; rather than meeting on hospital grounds (as with the other two committees), it rents space from a learned society. In terms of geographical proximity, its members come from a wider area than the LRECs, and it has, since its inception in 1997, always been a 'national' committee in that it has been able to review research from anywhere in the country. As the antithesis of a 'local' REC, we might expect such a committee to eschew the use of the knowledge drawing on institutionally situated or proximal-derived relationships; that it does not, that it goes so far (as we shall see) to artificially create situations that will *generate* local knowledge emphasises the role such knowledge plays in generic REC decision making. In turn, this highlights the centrality of trust decisions – where local knowledge plays the biggest role – in REC processes.

While Coastal MREC has always received applications from researchers with no institutional links, at the time of my fieldwork, this was not unknown for the LRECs I sat in on. Both

St Swithin's and Northmoor committees had recently become 'Type 3' committees – LRECs with the power to review research from any other site(s). This allowed members of these committees – such as Craig, an expert member of St Swithin's – to reflect on the differences between acting as an MREC and acting as an LREC:

> The fact that we sit occasionally as an MREC is something that I think perhaps has focussed our thinking because we may look at a protocol, and because we know the local individuals and we understand them and know how they're likely to behave in a given set of circumstances, be happy with something. But we have to be clear that we have to divorce the protocol from the individual if we're sitting as an MREC because in another place it might be outsourced to some complete lunatic who may not be able to deliver it at all and we obviously don't have the knowledge of those individuals.

While there are no obligations on researchers at particular sites to apply to those MRECs that meet close to them, given the importance of attending REC meetings to answer a committee's queries (see Chapter 3), applying to a committee that one can easily travel to makes sense. This in turn makes it more likely that members of the committee will work at the same institution as an applicant. This became clear in Coastal's review of a paediatric trial of antibiotics to treat a serious infectious condition associated with a specific immigrant community. The expert member assigned to this application provides an outline of the study (the design of which is largely unproblematic) and then moves on to the consent materials: 'But my main concern is the PIS are horrendous. They've tried to get them right but they've gone rather over the top. A mess.' Part of the problem is trying to cover too wide an age range – six year olds to young adults – using the same information sheet. There is clearly a need to provide PIS tailored to different age groups. Echoing the discussion in Chapter 1, question A68 is used as a trust proxy; however, this is not enough: 'They have tried, and in A68 they have everything but the kitchen sink in it, but it needs work.'

Yvonne, the lay member also assigned to this review, agrees that 'it's absolutely essential that this work is done since there is a dearth of treatments in this area and one can see why because clearly it's very hard to get right. Do they need to start as young as 6?'

At this point Colin, a lay member working within the NHS, begins to draw on his knowledge of the local area in question: 'I know the [patient] groups involved. Many of the families are people who are very concerned about their immigration status, so therefore the recruitment may be a problem.'

Yvonne: 'In A68 they do ask for our advice on quite a lot of things, don't they? There's a huge number of problems. Who is out there who could be involved and who could help?'

Again Colin draws on local knowledge, yet this time about the hospital this work is based at:

> Her boss has huge experience in this area and she could be pointed in that way. Could we advise them to look at how the clinic has dealt with these issues in the past? I have 'local knowledge'. I have confidence they have got, tucked away, the experience to do this properly.

The committee's Chair, Malcolm, brings matters to a close:

> I think it's better to reject this study, particularly as Colin could talk to them and advise them. I would not be happy to look at all the required changes on my own and would want it to come back to the committee as a whole.[56]

While the extensive changes to the consent materials needed means that this application would probably always have been rejected, local knowledge (in this case derived from institutional links) provides Coastal MREC with the reassurance both that the applicant could draw on research expertise at their hospital – 'Her boss has huge experience in this area' – and that a member of the committee is available to provide informal advice prior to resubmission.

But these kinds of cases, where members of the committee work alongside the researchers, are the exception. For Coastal MREC, local knowledge tends to come not from working alongside applicants in the same institution, but from researchers' regular (re)application to the committee, mimicking geographic proximity. More often than not, in a strong echo of Mark Jones and the St Swithin's LREC (or Parexel and Brent, for that matter), researchers choose to return to Coastal MREC for their reviews. For example, expert member Daniel starts a review by labelling the proposal: 'a Professor Gibbons study'. He then outlines a few minor problems, such as inconsistencies between various sections of the form (over how many sites the research will be carried out at, for example) before noting that 'We have accepted as satisfactory applications from this stable before.' As if to reinforce this, he notes that 'A68 sets out the ethical issues very well, including selection of names for recruitment prior to consent' although the PIS 'seems to indulge in dense verbosity, including a sentence with 57 words and no punctuation. It needs rewriting.' Colin (lay) is the second reviewer and agrees:

> I don't have much to add in that it's in line with the other applications we have seen from this stable which we have approved. The PIS does need better formatting. The introductory letter talks about 'this important study' and whenever I hear the word 'important' I think 'coercive'.

The study is given a provisional approval, pending revisions. Labelling this as 'a Professor Gibbons study' allows the REC to see this application in the context of previous work 'from this stable': generally trustworthy research in need of help with the PIS.

Yet there are other reasons for a specific applicant to re-apply to the same REC. Indeed, Coastal provides an example of an MREC specifically requiring future applications from the same programme of work to be submitted to it, rather than other committees. While this particular case is unusual, the study in question, the VFORCE study, first discussed in Chapter 1 – a research database drawing on NHS records – underlines the value of local knowledge, especially when members of a REC have trust concerns about applicants. When Coastal MREC reviewed the first application coming out of this study (around two years before my fieldwork) it was concerned that the long-term goals of this research were going to be extended beyond its narrow initial limits.[57] The database was to be constructed using anonymous information provided by clinics in specific disease groups. While patients would be informed of this research via adverts in the clinics in question, explicit prior consent for inclusion was not required. As a result, the committee:

> tried to build in fairly tough conditions against the misuse of material … we were concerned that people might have information about them being used to which they hadn't signed up to, because it's all very well putting a little advert in the bottom of the board in a clinic, who the hell's going to notice it? (Adam, lay member)

The second condition imposed was that all future VFORCE studies had to come to the Coastal MREC for approval, rather than any other committee, the basis of which was a fundamental lack of trust in the organisation running this research:

> I think the committee said 'we've looked quite carefully at the back-ground here. We know what kind of answers we're going to get. We know, if you like, what tricks people will pull' ... Now we've got quite an in-depth handle on what this organisation or what this research initiative is about. Therefore we're in a better position to protect the subjects of future studies if we're able to set that in that context. (Rob, expert member)

For members of this REC, it is clear that requiring repeat applications was in part because of uncertainty over what information might be given to other committees, especially with regard to the study's original aims. Put quite simply by Yvonne (lay), 'I think we're very suspicious of it. I really think it's as simple as that. I don't think we can trust them as far as we can throw them.'

These issues of mistrust came to a head in an application that planned on returning to the original patients enrolled on the database, to recruit them for an additional study, a form of re-contact that the original ethics application and approval had stated would not take place. The focus of this submission was to compare the effect of different incentives on response rates between different groups of patients, who were told that the study was about quite different research, involving a low-risk test. Leonard (lay member), who is leading the review of this application, suggests that this is a 'precedence setting study' since it involves re-contacting people enrolled on the VFORCE database, when normally the data is just used retrospectively. In ethical terms the primary aim of this study is to test the recruitment effects of inducements while *not* telling subjects what the purpose of the study is. The main issue is thus one of deception: 'The risk to participants is zero ...; the chances to benefits to the individual participants is zero but the population level benefit [of determining whether these inducements help re-cruit people into such studies] is clear.' But is this an acceptable way of doing epidemiology? There is then a debate amongst the committee over various aspects. Partly this is to do with a badly worded PIS, but mainly, as Yvonne puts it, 'It's simply wrong to pass a PIS that lies to patients – the PIS is based on a falsehood. What would the REC's ethical stance be in the future?' Discussion circles back to the original approval given to the VFORCE study, with the REC noting that the original posters put up in clinics did *not* talk about these sorts of studies, where data is sent back with patients' names on it: 'this will open the flood gates' to future re-contact studies. Colin, who is standing in as Chair for this meeting, agrees that this is a qualitatively different study from previous ones; the applicants cannot refer back to the previous studies approved by the committee and use those to support this application: 'This is an unconditional *no*.' The applicant researcher is invited in. Upon being told that this study has been rejected the researcher, without resistance, accepts that it is 'qualitatively' different from previous VFORCE applications – thus there is no dispute over the main point of concern for the committee. The researcher leaves and the committee appears satisfied.

For a number of members, this study – which takes a supposedly anonymised database and uses it to identify and contact patients – is a key example of why the committee asked for all VFORCE research to be referred to it. As one member put it:

> I think they were being disingenuous a lot of the time by saying, 'no, it's anonymised', and it wasn't anonymised, it was coded … at one stage they wanted to offer vouchers: 'hold on a minute, I thought they were meant to be anonymised'. And so there was mistrust. There's no getting away from it.

This point is reiterated by Adam (lay):

> because the rules that they originally agreed to, they would have agreed to almost anything to be allowed to get off the ground … and now of course they're established, they want to undo it. And they want to make different use of the data … they want to get back to the people whose data this is, and ask them different sorts of questions, and it was perfectly clear that that was what was going to happen to start with, but it's also perfectly clear that that's not what people have signed up to, when their clinic has gone in with this.

To some extent concerns about the previous limits placed on research using the VFORCE database allow the committee to make a clear decision about this particular project without having to resolve concerns around the deceptive nature of the proposed protocol. While research using deception is, of course, common in academic psychology, it is rare in the kind of research NHS RECs deal with.

A couple of months later, a re-application is made for this project. While changes have been made to some aspects of the proposed research, the key problems – deception around inducement and re-contact of patients – remain the same. As Yvonne points out, 'When did we last approve a piece of research where the researchers hid the purpose of the study from participants?' Leonard, who is lead reviewer sums up: 'This is a very full, but not very helpful response.' The applicant is invited into the meeting room; the Chair, Malcolm, thanks him for such a full response but says that there are a few points that the REC is still concerned with. Leonard leads the questions, asking about changes to the raison d'être of the VFORCE database (i.e. re-contact). The applicant is uncertain about what the problem is with this change of purpose: surely participants can opt not to take part in the study when approached? Leonard responds by making a comparison: that if you sign up to a service which will not send direct marketing mail, and you continue to receive such mail, then a contract/agreement has been broken. He then suggests that there is a case for setting up a VFORCE II database, where such approaching of patients is built into the original study design. But as things stand, to retrospectively change what the study is about is to violate the agreement with the patients. In the end, the committee rejects the application again.

On the face of it, Coastal MREC is applying a very high standard to this study. Changes to research once underway are not unknown, and the ethics review system acknowledges this, providing formal mechanisms for applicants to make amendments to their REC approvals. But the researchers in this case do not do this; they do not apply for a 'substantial amendment', presumably because for them, each application is a new research project. But the REC sees each application in the context of the broader programme of research (which they lack trust in) and the original restrictions laid down to control the use of the VFORCE database. Technically, the applicant is correct when he suggests that individual patients can opt not to take part in the study when approached; but this is to fail to understand the committee's concerns about this research and how ignoring this concern undermines trust in the applicant and their research. Thus the knowledge generated by repeat re-applications to Coastal MREC – re-applications required by the committee as part of the original ethics approval – allows the committee to make clear trust decisions about this research. Interestingly, in comparison with examples such as that of Mark Jones, this is less to do with the knowledge of the individual researcher and more a case of local knowledge about the project itself and the trust decisions about the company running the study.

However, somewhat paradoxically given the REC's initial scepticism, this regular re-application also has the effect of familiarising the committee with this programme of research, of generating understanding and trust. As a result, some members of Coastal suggest that there are advantages for researchers in such situations: 'I think in a way it's less frustrating if you're a researcher too because I think these studies are complex. I think the issues are complex and I think if a researcher were to get into another committee, they may have to re-invent the wheel.' This member, Colin, provides a neat summary of the dual value (to both committee and researcher) of repeat applications:

> it may be some of us saying 'we're keeping an eye on you' but I think it's really saying 'bring them back to us because otherwise you're going to be asked the same issues again and again and again' and I think every time we see a VFORCE database study in a way, we've got a template so we can review the study on its merits.

It is certainly the case that other research using the VFORCE database was passed by the committee with few changes required. This occurred even in those cases where members had concerns over the scientific quality of the proposed research (see Chapter 4). As Graham (expert member) suggests:

> I think that, because we've developed a certain, as it were, an expertise to understand the VFORCE database. So we're probably not as picky about it as we used to be. And [laughing], because we understood it better, and we've rehearsed the arguments often, repetitively, in the beginning, now we'll take it a little bit more on trust.

As already suggested, tensions around the trade-off between localism and centralisation are almost as old as the NHS itself; commentators on policy debates have tended to conclude that 'not all of the assumptions about how decentralisation can improve the delivery of health services are warranted', with one particularly cynical perspective suggesting 'much of the rhetoric about devolution [to the local level] ... can be seen as an attempt to devolve responsibility without power: to decentralise blame'.[58] Yet while most empirical research tends to highlight 'the striking dissonance between the political rhetoric and operational realities'[59] around the value of localism in healthcare, this analysis of the role of local knowledge in research ethics review tends to buck this trend, at least from the point of view of RECs and their members. Almost entirely ignored by these discussions, RECs underline how valuable knowledge arising out of institutional relationships and local connections with researchers can be to regulators and those they regulate.

As we have seen, historically, debates for and against localism have been used by policy makers to tactically protect the interests of medical researchers. In the 1980s, localism – the primacy of local REC decisions over an ignorant, faceless central committee – was the defence offered against the proposals of the British Medical Association's CEC, who were seen as lacking obvious research focus. Less than ten years later, localism – in the form of the wide variety of responses from RECs to the same multisite applications – was something to be overruled and avoided, albeit cautiously, through the use of MRECs. Yet beyond these contingent policy discourses, the structural features of the ethics review system reified the importance of local knowledge in REC decisions. Even changes put in place to reduce the impact of local autonomy provided some institutional spaces for localism to dominate; the provision for local revisions to MREC-approved consent forms and participant sheets, for example, or the changes breaking the local researcher–REC link, allowing applicants to apply to any REC in the country, but subsequently allowing opportunistic local relationships to develop, such as that between Parexel and the Brent committee. While, on a case-by-case basis, individual RECs may not develop localised trust relationships with applicants, a key insight is that the system is structured to facilitate this kind of trust-generating engagement. As Steven Shapin notes when looking at the trust-based relationships between venture capitalists (VCs) and the scientist-entrepreneurs they fund, 'If not every VC manages uncertainty through the resources of familiarity and the search for entrepreneurs' moral makeup, then the VC process itself manifestly does.'[60]

In terms of the questions posed at the beginning of the chapter – what does it mean for a regulatory body to be 'local', how do 'local' concerns influence the kinds of trust decisions these groups reach, and how do committee members see their role in the context of the local? – this chapter unpicks the twin aspects of the 'local' in this context. Both the roles of the institution and of local geographic proximity have historical roots in the policy debates over local review and centralisation, emphasising the importance of local knowledge to RECs and their trust decisions. Knowing researchers (either through working alongside them or from their repeat applications to the REC) helps frame members' questions and sets the context for their final decisions. Local knowledge can make or break an application. This is true (perhaps paradoxically) even for MRECs, multi-centre RECs which, after all, are not meant to have local affiliations or institutional connections. Even here, local knowledge of an applicant matters, to the extent that, in cases of low trust, a committee may artificially develop local knowledge, requiring re-application, as both a means of social control and (perhaps unintentionally) a trust-building mechanism.[61]

A key aspect of local knowledge of a researcher, of course, comes from meeting them face to face. While for some REC members this occurs in their roles as colleagues, for many, the only time they meet the researchers they review is in the REC meeting itself, when applicants are asked to attend and answer questions about their work. It is to this aspect of REC practice that the next chapter turns.

Notes

1 Department for Communities and Local Government, *A Plain English Guide to the Localism Act* (London: DCLG, 2011), 4.

2 For a review and discussion of localisation policies in the NHS, see: Mark Exworthy and Francesca Frosini, 'Room for manoeuvre?: Explaining local autonomy in the English National Health Service', *Health Policy*, 86:2/3 (2008), 204–212; Pauline Allen, 'New localism in the English National Health Service: What is it for?', *Health Policy*, 79 (2006), 244–252; Timothy Milewa, Justine Valentine, and Michael Calnan, 'Managerialism and active citizenship in Britain's reformed health service: Power and community in an era of decentralisation', *Social Science and Medicine*, 47:4 (1998), 507–517.

3 For details on the origins and early development of RECs, see Adam
   Hedgecoe, ' "A form of Practical Machinery": The origins of Research
   Ethics Committees in the UK: 1967–1972', *Medical History, 53*
   (2009), 331–350.
4 Department of Health and Social Security, *Supervision of the Ethics
   of Clinical Research Investigations and Fetal Research HSC(IS) 153*,
   (London: DHSS, June 1975), 2. The RCP report in question was:
   Royal College of Physicians, *Supervision of the Ethics of Clinical
   Research Investigations in Institutions* (London: Royal College of
   Physicians, 1973), which recommended that each committee should
   include 'a lay member … an individual who is not associated with
   the profession in any paramedical activity, i.e. a biochemist or a
   psychologist would not be considered as a layman for this purpose'.
5 Rudolph Klein and Janet Lewis, *The Politics of Consumer
   Representation: A Study of Community Health Councils* (London:
   Centre for Studies in Social Policy, 1976), 13 and 14.
6 UKNA, CAB 129/176/24, c(74)49, Secretary of State for Social
   Services, 'Democracy in the National Health Services', 21 May 1974.
7 The Working Group in Current Medical/Ethical Problems, Northern
   Regional Health Authority, 'Applications for ethical approval', *The
   Lancet,* 311:8055 (1978), 87–89; I.E. Thompson, K. French, K.M.
   Melia, K.M. Boyd, A.A. Templeton, and B. Potter, 'Research ethical
   committees in Scotland,' *British Medical Journal*, 282:6265 (1981),
   718–720; P.J. Lewis, 'The drawbacks of research ethics committees', *Journal of Medical Ethics*, 8:2 (1982), 61–64; P.A. Allen, W.E.
   Waters, and A.M. McGreen, 'Research Ethical Committees in 1981',
   Journal of the Royal College of Physicians, 17(1983), 96–98; Anon.
   'Editorial: Research Ethical Committees', *The Lancet*, 321:8332
   (1983), 1026.
8 Early debates over whether the RCP should set up a central REC can
   be found in: RCPA, MS4930 'Report of a meeting of Chairmen of
   Ethical Committees …', Thursday 17 October 1974.
9 Anon. 'Local ethical committees: Council approves revised report',
   British Medical Journal, 21 March, 282 (1981), 1010. Initial discus-
   sion of these issues can be found in: BMA Archive (hereafter BMAA)
   ETH49, 'Minutes of meeting', 20 October 1976, Central Ethical
   Committee. The discussion paper presented to the CEC is: BMAA,
   ETH49 'Local Ethical Committees: Discussion Paper', 5 March 1980,
   M.J.G. Thomas, Central Ethics Committee.
10 BMAA, Eth39 1983–84, 'Report of Meeting,' Wednesday 30
   November 1983, Central Ethics Committee. The earlier debates re-
   sulted in S.G. Owen of the MRC noting that: 'The question of a cen-
   tral national committee has been mooted in Council circles from time

to time over the years, and the view has always been taken that the disadvantages of such a body would predominate over any possible usefulness … we have feared a tendency to automatic and upwards reference in order that local committees might protect themselves'. Douglas Black of the RCP agreed: 'I would myself not be in favour of a national medical ethical committee as I do not see how it would override local ethical committees, to whom the applicants are presumably well known.' When the matter was discussed by a meeting of Chairs of RECs, it was rejected: 'There was no support among Chairmen for a national committee': RCPA, MS4932, Letter from S.G. Owen (MRC) to Sir Douglas Black, 3 May 1978; RCPA, MS4932, Letter (draft) from Douglas Black (RCP) to J.L. Gowans (MRC), 3 April 1978; RCPA, MS4930, Minutes of a meeting of Chairmen of Ethical Committees …, 26 September 1978.

11 For the statement of BMA Council's approval see: BMA 'Annual report of Council: 1983–1984', *British Medical Journal*, 288:6422 (1984), 25. The CEC's 'wishlist' of organisations to contact is in: BMAA, Eth85 1983–84, 'Discussion Paper: Local Ethical Research Committee and National (UK) Committee: BMA Report 1984,' n.d., Central Ethics Committee.

12 BMAA, Eth7 1985–1986, 'Note of the meeting with the Association of Community Health Councils …,' Friday 26 July 1985, Central Ethics Committee.

13 UKNA, FD7/3273, Letter (and supporting material) from J.D.J. Harvard to Sir James Gowans, 14 February 1985.

14 Rudolph Klein, *The New Politics of the NHS* (Harlow: Prentice Hall, 2001, fourth edition), 104.

15 UKNA, FD7/3273, Letter (and supporting material) from J.D.J. Harvard to Sir James Gowans, 14 February 1985.

16 UKNA, FD7/3273, Note for file from MPWG [Malcolm Godfrey, Gowans' deputy with a specific remit regarding the MRC's ethical matters], 13 March 1985.

17 UKNA, FD7/3273, 'British Medical Association and Ethical Committees', MPWG to Sir James [Gowans], 7/3/85. The RCP report was: Royal College of Physicians, *Guidelines on the Practice of Ethics Committees in Medical Research* (London: Royal College of Physicians of London, 1984).

18 UKNA, FD7/3273, Note for file by 'MPWG', 9 October 1985.

19 UKNA, FD7/3273, Malcolm Godfrey to J.A.D. Cropp, 26 November 1985.

20 BMAA, Eth26 1985–86, 'Minutes of CEC Meeting,' Wednesday 4 December 1985.

21 BMAA, Eth 34 1985–86, 'Minutes of CEC Meeting,' 19 February 1986.

22 Anon. 'Approval of ethical research committees delayed,' *British Medical Journal*, 292 (1986), 779.

23 UKNA, FD7/3273, Anon. 'More talks on research committee,' *BMA News Review*, April 1986.

24 BMAA, Eth11 1986–87, 'Note of the Meeting between the BMA and the Conference of Royal Colleges and their Faculties,' 9 September 1986.

25 BMA 'Annual Report of Council 1987–1988,' *British Medical Journal*, 296: 6626 (1988), 31. The revised RCP guidelines: Royal College of Physicians, *Guidelines on the Practice of Ethics Committees in Medical Research Involving Human Subjects* (London: Royal College of Physicians of London, 1990, second edition). For a more public – if equally overlooked – call for centralisation, see M. Warnock, 'A national ethics committee,' *British Medical Journal*, 24–31 December 297 (1988), 1626–1627.

26 For an early assessment of the Red Book, see J.V. McHale, 'Guidelines for medical research – some ethical and legal problems,' *Medical Law Review*, 1 (1993), 160–185.

27 Department of Health, *Local Research Ethics Committees* (London: HM Stationery Office, 1991), 10.

28 Donald Evans, Martyn Evans, David Greaves, Derek Morgan, and Neil Pickering, *Report to the Department of Health on the Conduct of Ethical Review of Multi-location Research Involving Human Subjects* (Swansea: Centre for Philosophy and Health Care, University College, 1992); Donald Evans, Martyn Evans, David Greaves, and Derek Morgan, *Trainer's Manual: for the Training of Members of Research Ethics Committees* (Swansea: Centre for Philosophy and Health Care, University College, 1992); Leigh & Baron Consulting and Christie Associates, *Standards for Local Research Ethics Committees. A Framework for Ethical Review* (Bristol: McKenna, NHS Department of Health Training Committee, 1994).

29 Clive Marritt, 'Local Research Ethics Committees: A View from the Department of Health,' in Bryony Close, R. Coomber, Anthony Hubbard, and John Illingworth (eds.), *Volunteers in Research and Testing* (Boca Raton, FL: Taylor and Francis, 1997), 53–58. Complaints about multi-centre research tended to follow the format of repeat applications to a number of different committees (usually as part of a piece of medical research) producing evidence of variability: Ursula Harries, Peter Fentem, William Tuxworth, and Gerald Hoinville, 'Local research ethics committees: Widely differing responses to a national survey protocol', *Journal of the Royal College of Physicians of London*, March/April, 28:2 (1994), 150–154; Paul Garfield, 'Cross district comparison of applications to research ethics committees', *British Medical Journal*, 9 September, 311:7006 (1995), 660–661;

Claire Middle, Ann Johnson, Tracey Petty, Lois Sims, and Alison Macfarlane, 'Ethics approval for a national postal survey: Recent experience,' *British Medical Journal,* 9 September, 311:7006 (1995), 659–660; Ala'Eldin Ahmed and Karl Nicholson, 'Delays and diversity in the practice of local research ethics committees,' *Journal of Medical Ethics,* 22 (1996), 263–266; Alison White, 'Research ethics committees at work: The experience of one multi-location study', *Journal of Medical Ethics,* 22 (1996), 352–355; M.E. Redshaw, A. Harris, and J.D. Baum, 'Research ethics committee audit: Differences between committees,' *Journal of Medical Ethics,* 22 (1996), 78–82.

30 George Alberti, 'Editorial: Local research ethics committees,' *British Medical Journal,* 311 (1995), 640.

31 An informal survey of LRECs carried out in the early 1990s suggested that industry-initiated multi-centre studies made up 14% of LREC workload: J. Moran, 'Local research ethics committees: Report of the 2nd National Conference,' *Journal of the Royal College of Physicians of London,* 26:4 (1992), 423–431.

32 Such commentators predicted 'considerable hostility to the use of a central committee by many local research ethics committees': M. Watling and J.K. Dewhurst, 'Letter: Central ethics committee might have to face hostile locals,' *British Medical Journal,* 311:7019 (1995), 1571. See also: Claire Foster, 'Will LRECS accept MRECs?' *Bulletin of Medical Ethics,* 133 (1997), 15–16.

33 NHS Executive, *Ethics Committee Review of Multi-centre Research: Establishment of Multi-centre Research Ethics Committees* (London: Department of Health, 1997), 2 (emphasis in original).

34 Richard Ashcroft, 'The ethics and governance of medical research: What does regulation have to do with morality?' *New Review of Bioethics,* 1:1 (2003), 54.

As other research has highlighted, members of LRECs articulated their concerns about MREC review in strong, moral terms, with Sarah Dyer pointing out that some of her interviewees felt that: 'Because of the "closer relationship" between an LREC and local researchers ... local review was more exacting.' Sarah Dyer, 'Applying Bioethics: Local Research Ethics Committees and their Ethical Regulation of Medical Research' (PhD thesis, University of London, 2005), 172.

35 'Review of Ethics of Multi-centre Trials,' CMO's Consultative Group on Research Ethics, RE/95/1, February 1995.

36 Pauline Allen, 'New localism in the English National Health Service: What is it for?'

37 Priscilla Alderson to Clive Marritt, 24 October 1995, CMO's Consultative Group on Research Ethics.

38 Rustam Al-Shahi and Charles Warlow, 'Ethical review of a multicentre study in Scotland: A weighty problem,' *Journal of the Royal College of Physicians of London*, 33:6 (1999), 549–552; Nick Dunn, A. Arscott, and R.D. Mann, 'Costs of seeking ethics approval before and after the introduction of multicentre research ethics committees,' *Journal of the Royal Society of Medicine*, 93 (2000), 511–512; Julia Lewis, Susan Tomkins, and Julian Sampson, 'Ethical approval for research involving geographically dispersed subjects: Unsuitability of the UK MREC/LREC system and relevance to uncommon genetic disorders,' *Journal of Medical Ethics*, 27 (2001), 347–351; Jat Sandhu and Mona Okasha, 'Letter: Evolutionary ethics – a continuing frustration?', *British Journal of General Practice*, 51 (2001), 674–675.

39 Richard Ashcroft, 'One year on: LRECs and MRECs in the West Country,' *Bulletin of Medical Ethics*, 143 (1998), 8–11.

40 Accurate at August 2020.

41 Adam Hedgecoe, Fatima Carvalho, Peter Lobmayer, and Fredrik Rakar, 'Research Ethics Committees in Europe: Implementing the Directive, respecting diversity', *Journal of Medical Ethics*, 32 (2006), 483–486.

42 Sarah Dyer, 'Applying Bioethics'.

43 Ryan Spellecy and Thomas May, 'More than cheating: Deception, IRB shopping, and the normative legitimacy of IRBs,' *Journal of Law and Medical Ethics*, 40:4 (2012), 990. See also: David Forster, 'Independent Institutional Review Boards,' *Seton Hall Law Review*, 32:3 (2002), 513–523; Barbara Evans, 'Inconsistent regulatory protection under the U.S. common rule,' *Cambridge Quarterly of Healthcare Ethics*, 13 (2004), 366–379.

44 For a critical perspective on the role of local issues in REC decision making, see: P. Wainwright and J. Saunders, 'What are local issues? The problem of the local review of research', *Journal of Medical Ethics*, 30 (2004), 313–317.

45 At the time of fieldwork, telephone boxes in UK cities were used as locations for cards advertising sex work. See: Phil Hubbard, 'Maintaining family values? Cleansing the streets of sex advertising,' *Area*, 34 (2002), 353–360.

46 Aggregating data from annual reports between 2002 and 2007, St Swithin's REC had a mean membership of 20. Of its 14 expert members, 12 (86%) worked for the hospital or its affiliated medical school. Of its six lay members, three had identifiable links with the hospital/medical school (either working for it or previously serving on a committee). This gives a mean 75% membership with links to the institution(s) the REC is associated with.

47 One aspect of the standardisation of REC processes is that each meeting is limited in terms of the numbers of applications – the 'slots' – that can be dealt with, usually around ten. If one REC's slots are full then an application is likely to be sent to another (probably nearby) committee.

48 Sarah Dyer, 'Applying Bioethics', 180.

49 Stephen Shapin, *A Social History of Truth: Civility and Science in Seventeenth-Century England.* (Chicago, IL: University of Chicago Press, 1994); Stephen Shapin, 'Trust, Honesty and the Authority of Science,' in R. Bulger, E. Bobby, and H Fineberg (eds.), *Society's Choices: Social and Ethical Decision Making in Biomedicine* (Washington, DC: National Academy Press, 1995), 388–408.

50 The classic statement of interactional expertise can be found in: Harry Collins and Robert Evans, 'The third wave of science studies: Studies of expertise and experience', *Social Studies of Science,* 32:2 (2002), 235–296. A more developed explanation is offered in: Harry Collins and Robert Evans, *Rethinking Expertise* (Chicago, IL: The University of Chicago Press, 2007).

51 In similar terms, Colin, a longstanding lay member of Coastal MREC, reflected on his own experience at an LREC which generated similar confidence on the part of researchers: '[LREC name] is a main REC, when people come to us, Professor Smith said "I wouldn't have wanted to go to anywhere else because you've reviewed the original stuff and you know what the issues are" … Applicants would rather wait, and I don't think it is, as the cynics say, it's because they know you're a pushover.'

52 For conflicts of interest and IRBs, see: Eric Campbell, Joel Weissman, Christine Vogeli, Brian Clarridge, Melissa Abraham, Jessica Marder, and Greg Koski, 'Financial relationships between institutional review board members and industry', *New England Journal of Medicine,* 355 (2006), 2321–2329; Robert Klitzman, ' "Members of the same club": Challenges and decisions faced by US IRBs in identifying and managing conflicts of interest', *PLoS One,* 6 (2011), e22796. For discussion of 'institutional reputation protection', see: Yvonna Lincoln and William Tierney, 'Qualitative research and institutional review boards', *Qualitative Inquiry,* 10 (2004), 219–234; Jonathan Moss, 'If institutional review boards were declared unconstitutional, they would have to be reinvented,' *Northwestern University Law Review,* 101:2(2007), 801–808. This latter aspect was mentioned by only one member of St Swithin's:

> I guess we could almost see ourselves as … also partly as a sort of research committee sort of saying, 'is this research worthy of going on at St

Swithin's, you know? Is this the kind of research that St Swithin's should be doing in a way? … Lay members can be caught up. 'I'm part of St Swithin's. I'm part of this great institution' as it were and you can go native … And there are times when discussions talk about what we're going to do very much with an eye on how is St Swithin's going to be perceived, which is an interesting sort of way to be thinking because why should we care how St Swithin's is perceived as it were? You know, there is a concern over the PR [public relations] aspect of the hospital coming from within our committee, which probably wouldn't happen with an MREC looking at some research in Dundee or something.

In terms of my fieldwork there was little other evidence of these sorts of 'PR' decisions.

53 N. Calder, M. Boyce, J. Posner, and D. Sciberras, 'Clinical pharmacology studies in UK Phase I units: an AHPPI survey 1999–2000', *British Journal of Clinical Pharmacology*, 57:1 (2004), 76–79; Jill Fisher, 'Co-ordinating "ethical" clinical trials: The role of research coordinators in the contract research industry,' *Sociology of Health and Illness*, 28:6 (2006), 678–694; Phillip Mirowski and Robert Van Horn, 'The contract research organization and the commercialization of scientific research,' *Social Studies of Science*, 35 (2005), 503–548.

54 Compare with, for example, the lay chair of one committee telling Sarah Dyer that: 'he depended on members of his committee knowing the local researcher in order to decide whether he or she was competent to carry out the research'. Sarah Dyer, 'Applying Bioethics', 178.

55 This sort of potential conflict of interest is almost inevitable in a REC that draws members from an institution that it also regularly reviews applications from. What *actually* counts as a conflict of interest is harder to define. The rules governing RECs make clear that research involving a member of a REC as an applicant cannot be reviewed by that committee, easily removing the most obvious examples where conflict of interest might occur. However, working with an applicant is not as clear cut. The same rules state: 'a member should withdraw from the meeting for the discussion and decision procedure concerning an application where there arises a conflict of interest; the conflict of interest should be indicated to the Chair prior to the review of the application, and recorded in the minutes': Department of Health, *Governance Arrangements for NHS Research Ethics Committees* (London: Department of Health, 2001), 27–28. While, on a number of occasions, I observed members raising potential conflicts with committee chairs, at no time was any member required to leave a meeting.

56 NHS RECs operate to a 40-day time limit for decisions, which leads committees to opt to reject those applications where a great many changes – too many to be dealt with by sub-committee or chairman's action – are required and need to be checked. In these cases, applications are rejected, 'stopping the clock' and allowing the application to return to the whole committee without running the risk of hitting the time limit.

57 As noted in Chapter 1, an additional source of the REC's concern about the VFORCE database was that it is commercially funded, yet seems to 'exploit' an NHS resource, tapping into RECs' deep-seated scepticism about commercial research.

58 Pauline Allen, 'New localism in the English National Health Service: What is it for?', 248 and 250.

59 Timothy Milewa, J. Valentine, and Michael Calnan, 'Managerialism and active citizenship in Britain's reformed health service: power and community in an era of decentralisation', 514.

60 Steven Shapin, *The Scientific Life: A Moral History of a Late Modern Vocation* (Chicago, IL: Chicago University Press, 2008), 299.

61 Other ethnographic research has stressed the way in which local knowledge and relationships structure prior ethical review: 'Even in a quite formal, bureaucratic REC, interpersonal trust and knowledge of local peculiarities played an important role in the ethical review process. Trustworthiness is not only judged on professional/expert reputation, but also on the basis of a REC's previous experiences with a researcher and his or her ability to deal with the REC's criticisms': Patricia Jaspers, Rob Houtepen, and Klasien Horstman, 'Ethical review: Standardizing procedures and local shaping of ethical review practices', *Social Science & Medicine,* 98 (2013), 316.

3

Facework, interaction, and the performance of trustworthiness

'The essence of IRB [Institutional Review Board] function is inter-action. IRBs are not nameless, faceless bureaucracies ... They know and are known to investigators and they spend a lot of time talking with them.'

Robert J. Levine[1]

'And the best way to find out who was evil, who was an asshole, and who was dumb was to spend some time with them.'

Steven Shapin[2]

Coastal Multi-centre Research Ethics Committee (MREC) considers an early-stage trial of a drug developed to treat advanced cancer. Things are not going well for the application. There are contradictions between the form and the participant information sheet (PIS); there are failures to complete the form properly ('they talk about "subjects" rather than participants ... there are apparently no comments in A68. There is a need to rewrite PIS in [plain] English': Adam); the consent boundaries are unreasonably broad ('there's very wide consent at the end of the PIS, where blood can be used in "any additional scientific research" ': Heather); and there is a clear lack of care and attention ('they will pre-treat with anti-histamines those patients who will have an allergic reaction to the drug – so they're either very clever or clairvoyant!': Malcolm). The committee ask the applicant who has been waiting outside to come in, to discuss their concerns.

Two applicants enter the room and sit down. The Chair informs them that a number of points will be raised by different members of the REC and Graham, an expert member, starts by exploring the earlier studies that have led to this application – the REC is concerned that the previous work has been conducted on solid tumours, yet the current application is on skin cancer. When one applicant responds that 'The majority of work was on melanomas actually', Malcolm replies: 'but the papers here only mention a solid tumour study where one participant had melanoma where there was a response'. It appears as if the way the application has been written is confusing (one applicant even suggests it might be a 'misprint'). As the discussion progresses the lead applicant apologises: 'I'm sorry that's confusing', leading one REC member, to mutter 'that's misleading...'.

Discussions over the design of the study – it is a complex trial with different arms and different doses – fare even worse. Adam questions why only one of the dose arms is run in combination with chemotherapy. The applicants' response – 'the investigators prefer that' – produces an exasperated reaction: 'I'm not interested in preference – they might prefer it, but I want to know what the rationale is!' The applicants' attempts to justify this choice result in Adam noting that 'I'm not saying you should have decided something else, I just want to understand the rationale.' The final response from the applicant appears unsatisfactory to Adam who, at the Chair's request, agrees to halt this line of questioning.

And so it goes on. In trying to reach a clear answer about whether the patients recruited into the study are chemotherapy-naïve and have never received treatment despite the advanced stage of their disease, Malcolm, the Chair, pushes the applicants. In return they fudge their answer, leading Malcolm, normally rather phlegmatic, to pointedly respond: 'Could you just answer the question I have asked?'

Finally the applicants leave. Malcolm sits back with an exhausted sigh: 'Goodness me, I feel more like I do at the end of a whole meeting.' Colin, the MREC's Vice Chair, reading the committee's mood, sums up the situation: 'It's a no go, isn't it? Resubmit.' Caroline, the committee's administrator, ever practically minded, highlights how long the discussion with these applicants has taken: 'I have apologised to people outside the meeting. Can we wrap this up quickly?' As the MREC moves on to the next application, Colin turns to Caroline and refers to the lead applicant's performance: 'He didn't do himself any favours, did he?' She agrees, noting that the time taken up with such applications means that 'If you get a bad study at the beginning, that's it! The whole meeting gone!'

Although the committee raised a number of serious issues about this application prior to the applicants entering the room, as we will see in this chapter, the researchers could have turned things round. As REC members noted: 'There are quite a few cases whereby we've been on the verge of rejecting an application and the applicant has come in and has persuaded us otherwise' (Malcolm, Chair) or a committee 'can be reviewing a study and you think, oh, this is going to be rejected. They bring the researcher in, and it's "oh no, is that what you mean?" And it can completely turn it round' (Caroline, administrator). The applicants' response to the REC's questions was crucial in determining the committee's final decision. This particular case highlights several aspects of how what we might call an applicant's presentation of self can influence the outcome of a REC's decision, how lack of clarity can swiftly become seen as a lack of honesty – 'that's confusing' becomes 'that's misleading' – how the failure to answer questions in a suitably clear way leads to frustration on the part of committee members, and how applicant attendance (or not) to answer questions shapes the timing and mood of a REC meeting, and the outcomes for specific applications.

The value of applicants being invited to attend and answer questions at a REC meeting is commonly mentioned when members reflect on the REC process. For example, following discussion of an application, Ben, the St Swithin's administrator, pops out of the meeting to see if the researcher in question has arrived; she is not there, and the committee waits for a bit, covering some other business. She still does not arrive, at which point Howard, the Chair, suggests that, in the light of the committee's discussion, the application be rejected and any re-application go before the whole committee, ruefully pointing out that 'if she had attended, if we'd seen the whites of her eyes, then we might have been able to clarify some of these points'. In conversation, while the more hyperbolic supporters of such attendance – 'It's absolutely invaluable to have the, to have them along. It's virtually impossible to make a decision based on a form without having a researcher along' (Damien, Northmoor) – were rare, the majority of REC members I spoke to felt that asking applicants to attend a meeting was an important part of the review process. More often than not, the value of this was couched in terms of how face-to-face communication allows one to clarify misunderstandings. Thus asking applicants to attend is 'useful in saving time when something hasn't been clear so that you don't have to go through a rejection process when all the while there was an answer but they just haven't been clear' (Celia, St Swithin's). As Craig, another member of St Swithin's REC, put it:

> you kind of short circuit a lot of misunderstandings with a face to face interchange because we're still carbon rather than a silicon-based organism and actually just seeing the individual and asking them specific questions can actually get through a lot of misunderstanding.

At the same time, some members also described the value of meeting applicants in terms that went beyond simply seeing attendance as a high-bandwidth means of clarifying uncertain technical aspects of applications. One member of St Swithin's described it as a way of gathering 'sort of meta-information, information that isn't in the application' which can be seen as 'useful but indefinable stuff about the investigators themselves' while Damien, an expert member from Northmoor, put it slightly more pithily: 'It's important to get an idea of the researcher themselves ... They may be exceptionally good at English and filling out forms but might be a complete muppet when it comes to research.'[3]

The importance of applicant attendance to NHS REC decision making is all the more surprising given both that it is comparatively rare as a feature of ethics review when compared to other countries' review processes and that it is relatively new as an aspect of standardised REC practice in the National Health Service (NHS). Compared to other countries, the attendance of applicants in the committee room to answer questions about their application is an anomaly. Indeed, in the US IRB system – perhaps the most obvious comparator for UK NHS RECs – the invitation for researchers to attend meetings, always historically low, is now at around 9%, with 'IRBs today ... less likely than they were a few decades ago to invite investigators to the review of their protocols.'[4] Even in the UK, before the mid-2000s, actually meeting applicants face to face to ask them questions about their proposal was seen as a procedural eccentricity practised by a minority of committees, including a couple in London and some elsewhere in the country, such as Birmingham and Wales. For example, exploring REC practice in the 1980s and 1990s, the Griffiths report noted that 'Only one of the researchers questioned had ever been invited to meet the LREC

[Local Research Ethics Committee] to discuss a project ... it was unusual to invite researchers to explain their projects because of the limited amount of time available.'[5] However, for some more locally situated committees it was, as Graham (Coastal MREC, expert) pointed out,

> a long-standing feature of REC practice ... If you're in a hospital, they're [i.e. researchers] all convenient. If you know them all, then you invite them to come. They come. They want their research to take place, and they know the only way they're going to get it done is to come. And they came. It's a pre-LREC practice.

In 1997, one of the new MRECs adopted the practice of inviting applicants to answer questions, despite resistance from some members of the new committee ('sort of whispered mutterings about this wasn't how everybody else works') and scepticism from the Chairs of the other MRECs, who saw it as 'completely impractical, impossible, something that couldn't be done, the meetings would go on for ten hours and so on'. According to one of the Chairs, it was only when the Medical Director of the ABPI – the Association of the British Pharmaceutical Industry – and the medical directors of two large pharmaceutical companies came along to one of the regular meetings that Chairs of MRECs held that the official view of applicant attendance changed. The industry representatives:

> grumbled that there was a certain amount of correspondence that went on and on, and why couldn't they come and talk to us? ... I think that was probably actually the pressure that made Central Office change its recommendations. So it did change at that point. I think there was just a groundswell of mutterings in the industry that the REC system is not terribly friendly towards the applicants.

Thus the main reason for the introduction of attendance as a standardised practice in NHS RECs was the perceived increased efficiency in information exchange and hence a way to clarify uncertainty and confusion about an application.[6]

This belief in the value of attendance as a means to clarity can be seen in Northmoor REC's review of an application deemed poor, shot through as it was with unexplained acronyms (Rose: 'I couldn't understand the acronyms'; Donald: 'Nor could I') and with a poor PIS (the eventual grounds for rejection). Upon realising that the applicant was not going to attend (Donald: 'Maybe they reread the application and thought better of turning up'), Rebecca noted the conventional view that 'It's unfortunate she couldn't attend since that could have cleared things up.' And of course sometimes the interaction between applicant and the committee can be seen in terms of a straightforward clarification of the issues and exchange of information. In one case, following an initial discussion of a proposal working on a chronic condition, St Swithin's REC invites the applicants in. The two researchers address most of the issues quite straightforwardly. The proposed restriction to patients who speak English is because patients will have already signed ten different consent forms and the applicants are reluctant to over-burden people without a good grasp of English. The use of technical language in the PIS (like 'gametes') was justified in that by this stage of their treatment, patients are used to this technical terminology and to change it for the research PIS would be odd. The applicants agreed that the PIS has some sections phrased in a coercive manner, accepting the REC's proposed changes. Overall the attendance was a success, with the Chair, Howard, summing up: 'It's good to have the researchers here to clear these things up.'

Members suggest that, for applicants, a clear benefit of attendance is when there is misunderstanding on the part of a REC (perhaps as a result of ambiguity in the original application) and speaking to the committee provides a chance to correct things. A good example is the case where Coastal MREC were considering an appeal – an application that had been rejected by another REC – where the application included an assessment of different surgical interventions in heart disease. While the rules around appeals mean that the application has to be the same as the one submitted to the original committee, the applicants have included a response to the previous opinion. Despite these written clarifications, the discussion is highly critical of the application. There are concerns about the study design (it is recruiting so few people that it should be classed as a pilot study) and its analytic approach – the committee's statistician notes that in terms of analysis the application simply says 'some statistical methods'. The issue becomes one of risk and benefit: given that one of the tests patients will undergo is a potentially harmful stress test, does the poor design of the study mean that it is not worth patients running the risk?

Pete (expert) suggests that there is a more fundamental problem in that these tests measure different things and can overestimate different factors. Since the disease being treated could have a wide range of causes, the applicants could thus, by focusing on the treatment in question, end up comparing people with what are in effect very different diseases. Following further discussion of the sample size proposed for this research, Adam, a lay member, seeks to clarify the REC's position with a question: 'is it the case that, if there is a concern over safety, it is better to have a larger number but proceed in a cautious way?' Gretchen, the committee's statistician, replies 'Yes'. Adam then suggests that, 'Given that the original committee was concerned and that that this resubmission has not covered their previous questions I think that we should reject.'

The applicants attend and begin by accepting the points about the size of the trial; if they could have, they would have renamed the trial as a pilot. They agree that this project will inform a larger study in a group that is quite hard to collect. During this discussion it is clear that there is a basic misunderstanding between the researchers and REC. The applicants, because they are cardiologists, assume a degree of background knowledge that the committee does not have. Contra the REC's expectations, this kind of surgery only happens with one kind of this disease, therefore there is no problem with multicausation. This makes a big difference to the REC's view of the application. The applicants offer a 'grovelling apology' – their own words – for the confusion. Colin suggests that 'the last 5 minutes have made a lot of difference to us', with other members of the committee agreeing. Following some discussion of the PIS, the applicants leave and the REC gives provisional approval.

Of course, in addition to attendance providing an opportunity for applicants to clarify what they want to do, it is also a chance to do the opposite, to confuse and perplex a committee. And it is in these cases where one can begin to see how an applicant's performance in front of a REC explicitly shapes the committee's decisions about trust. In another example from Coastal MREC, a study of treatment for specific tumours about which there are a number of typical concerns (consistency between PIS and application form and over-sampling tissue for genetic testing, for example), there is a major debate within the committee concerning the statistical level used by the Data Monitoring Board to decide whether to stop the research mid-trial. Leonard objects to the high *p*-value since it might mean that many more people in the placebo arm would miss out on treatment: if the *p*-value was dropped then more people might be swapped on to treatment. Gretchen, the committee's statistician, defends the use of a high *p*-value in the interim measure as an ethical addition to a normal fixed sample study with no interim, the point being that any interim measure is more ethical than not using one. During the attendance, the applicant's response increased uncertainty over this, with one member (Adam), complaining that, although he

> feels for both positions and I was initially persuaded by the applicant, he couldn't then give a ball park figure for Leonard's question *which is the sort of thing a PI* [principal investigator] *ought to know*. We ought to ask for clarity on this.

Here the concern is expressed in terms of basic competence – 'the sort of thing a PI ought to know', a kind of assessment which is multifaceted. At one level, making things clear to a REC is about demonstrating a particular set of skills which may come in handy when the researchers are actually carrying out their project. As Louise, an expert member from St Swithin's, put it: 'somebody who can't make it clear to us what about and why, makes you wonder about how clear they're going to make it to their [other] researchers or their patients'. More colloquially, 'If you get someone and you think "God, they're an idiot, they shouldn't be let loose on anyone", that can really scupper their chances.'

Yet more than this, sometimes the act of clarification raises more questions about an applicant's trustworthiness, most obviously by making clear that some important information has been left out of the written application, raising concerns (similar to those in Chapter 1) around inconsistencies within the written aspects of an application. For example, the St Swithin's LREC discuss a study aiming to improve the predictive tests offered to patients following surgery for a rare kind of tumour which interferes with hormone levels. This takes all patients undergoing such surgery, for any reason, and gives them low-dose dexamethasone suppression tests (a standard test) and, if they do not respond, taking them back into surgery. There is a discussion about the side effects of the drug used in the test, but apparently this is only the case in higher or longer doses, and there are some comments concerning the PIS, which does not follow the normal standardised format RECs tend to prefer. But overall, the REC seems well disposed towards this application. The applicant attends, and starts with an apology; in response to Howard's comment that this is a well-worded application, the applicant confesses: 'I have to apologise for the spelling mistakes – "pituitry" is embarrassing.' The discussion then proceeds in a straightforward manner, dealing with the timing of consent, whether taking the contraceptive pill is an exclusion criterion for women and the volume of blood involved in blood samples. However things take a different turn when the Chair asks about the drug administered to patients as part of the predictive test.

> **Howard:** 'What are the side effects at these doses?'
> **Applicant:** 'None.'
> **Howard:** 'So we can be confident in your statement that there are no side effects?'
> **Applicant:** 'It's not a treatment as such but rather it replaces nature.'

This comment produces a ripple of surprise around the REC.

> **Louise:** 'I hadn't realised that if you don't enter this study you get hydrocortisone and if you do you get dexamethasone.'
> **Howard:** 'The PIS needs to mention what is different from standard treatment.'

The applicant leaves. The REC is clearly unhappy to find out that the drug used in the test is not standard treatment as, they feel, is implied in the written application. Louise complains 'I feel a bit misled', before suggesting this applicant's behaviour might serve as an indicator of further problems: 'I'm a bit uncomfortable that there might be something else that he has left out?' Howard agrees. Abigail emphasises the unexpected nature of the revelation: 'That was a surprise'. And, as is often the case where a clarification has had an unexpected effect, other issues also become a problem, with Louise pointing out that she is:

> not happy with a PIS that requires people to stop the [contraceptive] pill beforehand. It's still ethical but it's a disappointing performance.

I'm not sure we want to congratulate them on their protocol now.
I'm not unhappy that they're swapping hydrocortisone with dexamethasone, *but I'm unhappy that he didn't think he ought to tell us*
(emphasis added).

Completing the movement from a simple lack of information to a
moral failing, Linda sums up: 'He decided that there were things
we didn't need to know', underlining how the information revealed
as a result of attending the REC meeting can raise questions about
an applicant's trustworthiness and character. The lack of clarity
around the replacement of hydrocortisone is interpreted, not in
terms of mistake but as deliberate deception: the language is of the
REC being misled rather than terms relating to competence (such
as 'confusion' or 'clarity').[7]

It is clear, therefore, that while successful attendance before a
REC is often based upon the performance of competence on the
part of the applicant, this performance is also a moral marker: how
the applicant responds to a committee's questions – both in the *content* of and the *tone* of the reply – acts as a marker for an applicant's
trustworthiness.[8] In terms of broader debates around trust, the
importance of meeting applicants is a version of what Anthony
Giddens (after Goffman) calls 'facework', 'trust relations which are
sustained by or expressed in social connections established in circumstances of copresence', a form of trust which balances the anonymous and impersonal nature of modern 'expert systems'.[9]

At the most basic level, facework depends upon *who* is sent along
to answer questions about an application, telling RECs something
important about how the research is viewed. As Heather, a member
of Coastal MREC, recalls, 'We had, last time, somebody who completely destroyed his case by being there. I was more confused when
he finished than when he started. He couldn't answer the question.' The key problem arises 'when people can't answer because
the right person isn't there, that's when people object, because they
look embarrassed and frustrated and they just can't answer. And
you think, well, somebody else should have been here answering
those questions.'

Another example of this came in summing up an application
where there are contradictions between the proposal, the PIS, and
what the researcher said were the aims in the actual meeting, as
well as a number of potential side effects that had not been spelled

out in the PIS; Coastal's Vice Chair, Colin, notes that 'The study's quite complex and the company sent along someone who is really quite junior, thinking that the study was a run of the mill study.' The point is not only that a more junior, inexperienced, member of staff may have difficulty dealing with all the questions that might arise but also that the act of sending junior staff along can be seen by the REC as an indicator of how seriously (or not) a PI takes ethics review.

If one key element to performing competence (and thus rendering yourself trustworthy) is the ability to answer technical questions authoritatively, then one solution might seem to be to bring more than one person to the meeting, to increase the chances that you will be able to answer questions. However, this is not always a good idea, as St Swithin's LREC demonstrated in their review of a late-stage clinical trial of a drug used to treat schizophrenia, used in an injectable rather than standard oral format. Since the REC has no psychiatric expertise, the Chair and administrator have organised an external review, which points out the main problem, which is that the trial did not involve a Data Safety Monitoring Board (DSMB) to oversee possible problems like higher suicide rates. Introducing the discussion, Howard suggests that 'We're happy that the drug's safe in itself', only for Craig to note, 'But the drug is being used in totally different stage of illness from its licensing conditions.' In discussion, the key issue that arises is that of consent, and patients' ability to give this for the research, given that they will be recruited when they have a relapse, and may have a reduced capacity for consent. The UK system includes the possibility of appointing legal representatives in cases of reduced capacity and the committee wonders if the applicants could look at the relevant guidelines.

Following discussion of the PIS, the applicants – the PI and a company representative – attend. As Chair, Howard starts by presenting himself as a 'lay person' with regard to the focus of this trial as well as mentioning the REC's lack of psychiatric expertise. The researchers share out the answers between them fairly evenly, but they have problems with a number of the REC's questions. With regard to the need for an independent DSMB the company representative does not know why one is not involved in this trial and asks to defer an answer until she can check up. In terms of consent, the PI draws on their apparent clinical experience – and, of course, the

REC's self-professed lack of experience – suggesting that people in the early stages of relapse would be able to give consent. Then Craig asks whether the application conforms to the regulations around consent, particularly with reference to legal representatives.[10] The applicants look blank and seem to have no idea what he is talking about. They agree that there is a need to re-consent people who re-cover under treatment and, when asked whether a seven-page PIS is appropriate for these sorts of patients, the PI replies that they would be 'talked through it'. The REC looks unconvinced.

The applicants leave the room. The REC is unimpressed. Partly the problem is that, while the applicants were in the room, some members of the committee have looked through the PI's CV (attached, as per usual, to the application). Members now raise issues not discussed before the applicants attended (although of course the written information available to the committee has not changed), suggesting that the PI has very few publications under her belt and seem to just consult for industry projects. This is regarded sceptically. But there are also problems with the applicants' responses during the meeting, particularly over their lack of knowledge regarding consent issues. The poor performative aspects of the applicants' responses is underlined by Dean: 'When she starts an answers with "I would imagine…" '. Abigail follows up, noting that 'the PI was totally baffled by the protocol flowchart the representative showed at the back of the application'. The REC's critique begins to gather steam:

Samar: 'Their answers on a test case were just cobblers; we caught them on the hop with that question. I suspect this is driven by licensing.'
Celia: 'Exactly.'
Samar: 'There's nothing wrong with people coming to the committee and saying they're doing a licensing study, but they shouldn't cloud it by saying it's for patient benefit.'

Yet these concerns, effectively about applicants misleading the REC, are not presented as issues that could lead to a rejection of this application. The impression is that if these were the only issues then the committee would approve the trial.

Hilary: 'But there is real issue about consent.'
Howard: 'And whether we approve this study or not, they need to come back with a well thought out package on

> legal representatives ... Are we happy with the study, the underlying trial?'
>
> Tom: 'It's not for us to challenge the underlying science: it's a bit weird but that's their issue.'[11]
>
> Howard: 'We can "invite them" to reconsider the DSMB but the PIS is where there are real problems. This is a reject, but we expect to see a resubmission because of the way the clock ticks; the consent issues are too great, it'll have to come back to the whole committee.'

The kinds of changes that the REC is asking for – around the need for a DSMB or consent issues and the need for a legal representative – are not things that need necessarily result in a rejection. I observed other applications with the same kinds of problems getting a provisional approval. The determining factor is the applicant's performance. Surprise at the lack of a DSMB on one's own trial, or the possibility of legal representatives in cases where there might be limited capacity to consent, does not engender confidence on the part of a committee. Clearly, having more than one person attend will not necessarily mean that researchers make a more convincing case to a REC. Unsurprisingly, having two people performing badly does not increase one's apparent trustworthiness. In addition, how an applicant performs in front of the committee can send members back to the written application, to look more closely at, for example, a CV, to find things there – working only as a consultant for industry – that reinforce the REC's (unflattering) impression.

This case reinforces that a poor performance is not just about content (e.g. the lack of something) but also tone (*surprise* at the lack of something), a point brought home when Northmoor and district LREC consider a clinical trial of a new morning-after contraceptive pill (MAP). In discussion, the committee has a number of concerns – safety issues about the trial of the higher dose of the drug being tested and problems with the PIS (Rebecca: 'I thought the PIS was a bit long and badly structured. The purpose of the study needs to be closer to the beginning') – but the main one is that that the study, recruiting from a family planning clinic, will recruit women in the control arm (on the normal MAP) after 72 hours post-conception. Since this is later than the current MAP is licensed for, the REC is worried that women on the control arm will experience a higher

rate of failure than if they were given the normal post-72-hour option of a contraceptive coil. The applicant, a representative from the
company developing the drug, attends. What follows is what my
scribbled fieldnotes describe as a 'car crash'.

Martin
(Chair):
'This is an important area of research but there is
already an existing product on the market. One of
our big concerns was the plan to recruit women who
were beyond the 72-hours limit, who would be ran-
domised to either the new product or the existing
product. But these women would be deprived of the
best standard UK treatment, a coil, since the old MAP
is not licensed beyond 72 hours.'

Applicant: 'I think the informed consent is quite clear about
any standard of care.'

Martin: 'Well, that's another point, but we think it's unethical to deprive people of best treatment.'

Applicant: 'There are some studies of current MAP beyond 72
hours.'

Martin: 'But it was not effective. Did you consider just
limiting the study up to 72 hours? You are trying
to do two studies in one, or perhaps you're just recruiting in two bands (old MAP/new MAP) and a
post-72-hour band (new MAP/coil)?'

Applicant: 'Can I call the PI?'

Martin: 'Not with the time limit we work under.'

Applicant: 'OK. I guess the old MAP can be taken beyond 72
hours – it's standard practice in France and in the
UK, I guess.'

Rebecca intervenes: 'But it's *not licensed* beyond 72 hours, which
is the end point we have to take into account.' The discussion then
addresses issues around the safety data for the proposed higher
doses of the test drug, and the applicant reassures the committee
that the company have considerable information on this. Finally,
the discussion turns to the safety issues raised by recruiting women
from these specific clinics:

Martin: 'But we have some questions on recruitment: is
it your intention to get all your subjects from
Northmoor family planning clinics?'

Applicant: 'We would like to add some clinics from elsewhere in the UK.'

Martin: 'We were unhappy that GPs were not being informed. We understand issues around confidentiality but were concerned about possible side effects and the need for GPs to know. Can you comment?'

Applicant: 'The main issue is confidentiality.'

Martin: 'Sure, but if she, the participant, doesn't want to inform her GP then she opts out of the study: family planning clinics do not provide 24-hour care.'

Megan: 'It's also the nature of the treatment: young women may not tell their family that they have entered the trial and if there is a reaction and the GP does not know then how would anyone know?'

Applicant: 'If we put in a 24-hour contact number?'

Megan: 'But if she collapses then the family will phone the GP or A&E: there is no safety net in the system.'

Applicant: 'But the [family planning] clinics are aware of the trial.'

Megan: 'Yes, but how will *they* know if there has been a problem?'

The applicant leaves. Martin sums up: 'From the discussion we've had we have to reject.' Megan agrees: 'If they are really after safety data then why is there no safety net? If there was a death, how would they know?' The study is rejected.

On one level what we have here is a clash between legitimate concerns. The REC is worried that, by not informing the participant's GP there is an increased risk, since if there is an adverse reaction to the drug no one treating the participant (their GP, the A&E they might end up in, or their family) will know they are on a clinical trial. The applicant's position is that they wish to respect the participant's confidentiality (given the sensitive nature of the drug) and that informing the GP might compromise this (and, of course, participants may not want to take part in a trial if they know their GP will be informed). Yet this is against the background of the previous discussion around the design of the trial and the recruitment of women beyond the 72-hour post-conception point. The REC could approve the study on the condition that specific changes were made (for example, no post-72-hour comparison arm with standard MAP, only coil; GPs to be informed of trial participation); however, the applicant's performance – a mixture of uncertainty and intransigence – makes this unlikely. An applicant who asks if they can call the PI to clarify a point of clinical practice (use of the standard MAP beyond 72 hours) does not inspire confidence, especially if the conclusion is that '*I guess* the old MAP can be taken beyond 72 hours.'[12] A response to a question about the complexity of a consent form that runs 'I think the informed consent is quite clear about any standard of care' comes across as unhelpfully contradictory, as does the response to the licensed use of the current MAP: 'There are some studies of current MAP beyond 72 hours.' Thus in this case, the *tone* of the applicant's performance seems to swing the committee's decision against simply approving with amendments. In this context, as Sarah Dyer has noted, in meeting the applicants it

> is not so much that committees understand the research any more than they did from the application form and research protocol alone but that they feel they can trust the researcher who appears competent. In this sense confidence becomes a proxy for ethical.[13]

The clearly 'embodied' nature of such confident performances raises questions about possible 'demographic' variation – for example, by gender or race – and whether such embodied confidence is the preserve of white males, for example. On the basis of my fieldwork, the answer is, it is hard to say. The limited numbers of attendees from black or minority ethnic background mean that it is hard to draw conclusions about the role of race, although the predominantly white, British composition of RECs might lead us to consider this as a factor. While there were more women attendees, drawing conclusions about the 'gendered' nature of such performances is hard, not least because we know that the nature of modern medicine requires successful women to mimic many features of male behaviour.[14]

A clear tonal difference in performance can be found in a comparison of two similar applications reviewed by Costal MREC. The first is an international study of a compound to treat cardiovascular disease, the trial of which involves testing in conjunction with a statin, involving a total of over 900 participants, 85 of whom are from the UK. When the applicant attends the meeting, questions centre on the standardised approach to prescribing the statin, which might be different from a patient's clinical care and which is potentially risky. When pushed by Sarah (expert) as to 'why you couldn't recruit patients already on this particular statin, at those particular doses?', the PI replies: 'That would make recruitment hard' and the company representative talks about how patients are recruited globally and if the UK arm doesn't recruit fast enough then its quota could be 'mopped up' by other countries before UK recruitment could complete.

After the applicants leave, it is clear that justifying a potentially risky design on the basis of internal participant recruitment competition between different trial units is something of a faux pas. Pete was the first to speak, noting that 'I do not accept that you cannot recruit patients already on this drug.' There then followed a quick-fire discussion of the poverty of the applicant's performance:

Daniel:	'He shot himself in the foot, the more he went on.'
Malcolm (Chair):	'Before they came in I was minded to give a favourable opinion, but this went downhill rapidly.'
Sarah:	'Statins are good drugs but they have problems, and the claim about the need for "global recruitment" ...'
Malcolm:	'Sometimes when an applicant attends it make you much more inclined to accept; but in this case...'

The application is rejected, with the Chair noting that there will almost certainly be an appeal, so the REC needs to make clear its grounds for rejection. What underlines the crucial role of the applicant's performance in this case is that, around six months later, the MREC reviews a similar application – an industry study looking at a combination of already licensed drugs for hypertension – where a different decision comes about. The point is not that the REC is happy with the proposed recruitment of treatment naïve patients, far from it. But rather, that the applicant draws on a different set of ideas – about the feasibility of research design, for example – which makes it easier for the REC to trust him. The first question seeks to clarify the groups that the trial will recruit, pointing out that, with 'people who are clinically stable, it might not be suitable to take them off treatment to put them on the trial, might it?' The applicant responds positively – 'That's an excellent point' – making clear that 'the patient population is treatment-naïve or uncontrolled on single treatment'. There is then a discussion around the safe dosage level of the drugs in question (there are concerns that the trial will start patients on higher doses than necessary) before questions return to who will be recruited:

Sarah: 'Using treatment-naïve patients is inconsistent with the guidelines of the British Hypertension Society.'

Applicant: 'We are aware of that; it's not normal to start with two drugs, though it is standard in the US. Speaking to UK specialists, they say that they do test with combined doses.'

Sarah: 'Yes, but this trial uses GPs, who may not be as experienced as hypertension experts.'

Applicant: 'I understand that, but all I can do is reiterate that the combined drug at this dose is safe.'

Sarah: 'What would be the justification for GPs to go in with this study rather than the level they are guided to do? They are not following any formal clinical guidance. There's no rationale for GPs to put patients in, in the case of treatment-naïve patients.'

Applicant: 'We are trying to show that aggressive blood pressure treatment is of value to high-risk, treatment-naïve patients. This is evidence-based medicine. I accept that this is not following the current guidelines; we are working beyond them.'

Sarah: 'But how do you know that in treatment-naïve patients they couldn't be treated with lower doses? There is an ethical issue of over-treating.'

Applicant: 'We can't answer that.'

Malcolm: 'Which does present an ethical problem.'

Applicant: 'I accept that, but it would be very hard to write a protocol that would allow us to work on this group.'

Daniel: 'What would be the problem with just focusing on patients who can't be controlled and then moving on to treatment-naïve patients?'

Applicant: 'We think it's better to work on as broad a population as possible. I don't see such a distinction between "naïve patients uncontrolled" and "treated
patients uncontrolled".'

Daniel: 'In numbers, if we were to say it shouldn't be naïve
patients, would it harm the study? As a GP, I think
the number of naïve patients I might recruit to this
study would be very small.'

Applicant: 'I think the number of changes would be minimal.'

Malcolm: 'I think that's our solution, isn't it?'

The key point is that, while there is a discussion about the impact
of excluding treatment-naïve patients on recruitment numbers, this
is couched in language that we might see as *internal* to a scientific
discussion. With regard to safety, the applicant refers to current
clinical practice both in the US and (contra formal guidelines, to be
sure) the UK, making a clear claim – 'all I can do is reiterate that
the combined drug at this dose is safe' – about the safety of these
drugs at these doses. In terms of the possible efficacy of the combined drugs, the applicant refers to the concept of evidence-based
medicine and working beyond current guidelines, as well as offering
scepticism about specific categories ('I don't see such a distinction
between "naïve patients uncontrolled" and "treated patients uncontrolled"'). Crucially, unlike the applicant in the previous study,
this PI does not justify his recruitment choices by reference to the
internal political economy of trial recruitment that, while a genuine
concern for individuals working within industry, is not an appropriate set of values to display for an NHS REC. Indeed, in this
second case, by accident or design, it is not the applicant who raises
the explicit problem of recruitment for the trial – 'would it harm the
study?' – but a member of the committee itself. The end result seems
far from a complete victory for the applicant. After they leave, the
Chair offers a summing up – 'They seemed to accept that they could
only use treatment-unstable patients' – and there is a discussion
about the dosage levels which settles on a position which may well
be problematic for the trial: 'Are we saying that if they came back
and have made corrections, and at half the dose we would accept
it?' (Malcolm). The committee agrees that, provided these changes
are made, the application can proceed. While technically, in terms

of design, this study is much like the previous one, it is treated differently – it is not rejected, with the main difference lying in how the applicant responds to the committee's questions.

As should be clear by now, beyond any information exchanged, how applicants present themselves – their performance of trustworthiness – can shape the outcome of an application. Even in those cases where RECs do not decide to reject, attention is paid to an applicant's demeanour. In the broadest terms, REC members articulate this in rather vague, non-specific ways:

> And to some extent it's about getting a grip when you see people and it's almost not like questioning. It's just that feeling you had I suspect of, are they a good person to be running a trial and do we get the feeling they know what they're doing? That we're comfortable with them sort of approaching patients and everything will be right with the world because they seem to be a decent person or something (Ben, St Swithin's LREC).

More specifically, and perhaps unsurprisingly, off-hand or rude behaviour in front of a REC can lead to questions being asked about how suitable a researcher is. As one REC member put it in a lighthearted way: 'You sometimes think, "oh my God, they're a sociopath with no social skills". [laughs] You know, you really shouldn't be near patients, let alone running a trial.' In a slightly more considered way, Hilary, an expert member of St Swithin's, discusses an application:

> It was given an unfavourable opinion and then they attended again, and some of the things that they said were really quite outrageous in terms of people's sensibilities and their perception of what they felt people would be amenable to and wouldn't be. Again, you can't help but think, certainly with me, that some of your views there are very personal and perhaps you shouldn't, sort of rely too heavily on them. But sometimes when you meet people … some of the things that you were worried about, they explain and you felt quite happy with what they've said and sometimes it's just the fact that they've used terminology in a way that is unfamiliar to you, but obviously familiar to them and it just, everything slots into place and you feel comfortable that, 'yes, this is okay'. And then other times, when you perhaps had a bit of a concern, it just says, yeah, I'm right, I don't feel happy about this, and having seen them and heard what they've

got to say and hear their perspective of what they've written, I still don't feel happy about it, and these are the reasons why.

In this case Hilary, despite her caution about relying too heavily on views based on personal interaction, emphasises the way in which meeting applicants can reassure or reinforce concerns. This is often articulated in terms of 'arrogance'. Charlotte (Northmoor) recalls 'a couple of really cocky dentists' who were:

> young, arrogant, very bright ... but at the time, they were morons ... their bearing did work really badly against them and I think we might have actually passed a piece of quite poor-quality research because it wasn't dangerous or anything. Actually, we were all saying that they were so arrogant and they really should think again how they thought about people.[15]

One key point about such arrogance is a failure to acknowledge the need to revise one's application: 'We've had some terrible people, arrogant people attend, and they're not going to succeed, really, because they're so arrogant, they won't admit that the change is necessary' (Graham, Coastal MREC). And the issue of perceived receptivity to a REC's suggestions is crucial:

> Mostly people who come, I think, are usually receptive to suggestions. And quite a lot of people on the committee, I think, do their best to be as helpful as possible and suggest things that could improve the research, which I think is very important. If they're not very receptive, then you become a little concerned at times' (Rebecca, Northmoor).

We shouldn't assume, however, that RECs worry about applicants simply on the basis that they are rude or arrogant. Perhaps counterintuitively, even the most good-natured of performances can raise concerns: an applicant is invited into the meeting of St Swithin's LREC. What follows is a good-natured discussion, with the enthusiastic applicant accepting that an independent data monitoring committee would be a good idea. After the applicant leaves, the following exchange takes place:

Ben: 'Is he too gung-ho about the treatment?'
Dean: 'But how do you temper that?'
Ingrid: 'He just talked himself into trouble ...'

Howard: 'If only he'd stopped talking a couple of sentences
           before ...'
Dean: 'We can temper his enthusiasm by making sure that
           the PIS is as accurate as possible.'

Someone else points out that this would not reduce his enthusiasm when he discussed the study with potential participants, and the risk was that his personal charisma would override any misgivings included in the PIS. Beyond re-emphasising the central dilemma of prior approval – that whatever is agreed to in the REC meeting, the committee has no way of knowing how the applicant will behave when recruiting or treating research participants – this example also underlines how aspects of one's demeanour that we would normally approve of – enthusiasm for one's research – can lead to concerns on the part of a REC.

What comes though clearly from examination of applicants' performance is that RECs' assessment of the risk associated with particular proposals is far from the asocial calculative process so beloved of bioethical and medical texts.[16] Rather, decisions about the acceptability of research risk are intimately bound up with who is going to carry out the research and how they have performed in front of the committee, an – if you will – interactional model of risk assessment. 'Riskiness' is not a feature of just the written application, but also of how the applicant performs. Therefore, whether an application is risky or not is not a fixed characteristic of a piece of proposed research but rather is interactional. Thus, as we can see in the two examples that follow – applications that were reviewed by Northmoor LREC in consecutive months, with substantially the same members present – it is perfectly possible for two proposals, with very different levels of physical risk, but the same weaknesses, to be treated in quite unexpected ways depending upon the performances offered up. The first case, which involves using heart surgery to treat a specific form of migraine, involves considerable risks to participants. While there are some concerns about the need for a longer post-operation follow-up, to track the safety of the treatment and questions about the efficiency of the drugs used in the control arm, the main focus of the REC's comments is the PIS, with several members of the committee chipping in with their own concerns:

Martin 'The PIS is a bit of a mess, isn't it? There's a
(Chair): duplication of sections ("What are the side effects of
taking part?").'
Rose: 'It's a mess.'
Frank: 'I found the study title, as a lay reader, just weird.'
Martin: 'Yes, it's a title chosen to fit an acronym.'
George: 'I would like them to be more upfront about what they
are going to do. It's not until the bottom of the third
page that they tell you they will shove the device up an
artery in the thigh.'
Megan: 'They don't say that you might be allergic to the con-
trast material used in the operation.'
Neil: 'There's an X-ray exposure question …'

When the applicant attends, however, the PIS is not the main focus of discussion. This is often the case, since many of the issues around the wording of PIS are communicated to applicants in writing, to save time in the meeting. Instead, Martin, the Chair, pursues 'a few clarifying issues' around a previous study (appli-cant: 'We were in that. It's finished and data has been presented and now four or five other trials are up and running, funded by different companies') and the additional information that would result from the proposed research (applicant: 'For one, it's a different device and we suspect that some devices will have different effects from others, depending on the material the device is made from'). Martin points out that these details should be in the PIS:

Applicant: 'It's in there, isn't it?'
Martin: 'No.'
Applicant: [apologetically] 'Well, it was in the last PIS.'

Following reassurance about the effectiveness of the drugs being used in the control, and acceptance that the research team will need to be more proactive in staying in touch with partici-pants after the operation (in order to track safety data), the appli-cant leaves. Martin sums up: 'He managed to clarify most of the points. They need to do a lot of work on the PIS but apart from that, is this OK?' The applicant's performance before the com-mittee was largely neutral, answering technical questions clearly and confidently and responding to an error in the paperwork in a straightforward, though apologetic, manner. This was enough to reassure the committee regarding the risks involved in the proposed research. The numerous changes required to the PIS become a technical issue.

The second application is, on the face of it, lower risk: a healthy volunteer study looking at the possible impact of particular kinds of dietary carbohydrate on people at risk of type 2 diabetes. The main intervention is inclusion of the carbohydrate, or a placebo, in participants' diet and monitoring this via blood tests. It is these tests that the REC is initially concerned about:

Amanda: 'They're sticking needles in all over the place.'
Sally: 'There's a cannula in each arm and in one hand. They [the participants] have to come starved to the clinic and in 2–3 days they will have 1.4 litres of blood taken. They are then given the supplement and they have an extensive lengthy food diary and bowel habit diary which is not mentioned in the PIS. Overall the PIS did not contain enough of the sort of information it should.'
Amanda: 'They underplay risks – they don't mention infection from the cannulas.'
Charlotte: 'They do get given £200, which is fine I think, but do they get it if they drop out?'
Frank: 'Would any of you have volunteered?'
A chorus of 'no's runs round the table.
Charlotte: 'I wouldn't fill in the forms properly. I'd be too busy. But we shouldn't make assumptions about why people would take part.'
Megan: 'It's fine as long as they put everything in the PIS, which they don't. There is a problem with the technical nature of the language in the PIS but also the application form.'

The applicant attends and Martin starts by pointing out that the REC 'did feel this was a complicated enterprise for healthy volunteers'. The applicant's response – defensive, and, to some extent, evasive – sets the tone for this exchange: 'We've never had a recruitment problem with similar studies.'

Frank: 'We felt there are one or two points that dissuade us, that weren't in the PIS.'

Sally: 'The blood sample, it's not small is it?'

Applicant: 'It's a series of small samples adding up to a large volume.'

Sally: 'I don't think you mention the bowel diary in the PIS and it certainly needs to say how explicit it is. The PIS doesn't give enough information.'

Charlotte: 'Is the £200 only if they do the whole study?'

Applicant: 'It depends how much of the study they do. If they drop out halfway through then we would have to think about it. If they had a good reason, then we would pay them.'

Megan: 'Is there a risk of anaphylactic shock with the local anaesthetics?'

Applicant: 'We haven't mentioned it normally but we can.'

The applicant leaves and the REC begins its discussion. Sally starts with a straightforward suggestion: 'I would like to see the PIS again' but Charlotte phrases her concerns in terms of the applicants' moral character: 'they're a bit cavalier'. Returning to the technical issues, but with an important moral subtext, Megan asks, 'Are we happy to give a provisional approval to this study when so much is missing from the PIS?' What seems to finalise the decision is a moral/procedural 'one-two' with Frank articulating the REC's lack of trust in the applicant – 'In other words we don't think amendments are enough, we want to see the whole thing again' – and hence the need for the whole committee to review the revised PIS, and the committee's administrator, Jane, pointing out the timing issues around the 60-day limit for REC decisions and thus 'if it is to come back to the whole committee then it has to be a reject.' Frank: 'So a reject.'

Why was the second (lower-risk) study rejected on grounds (the extensive changes required to the PIS and the need for the entire committee to review these) that could have quite easily applied to the first, higher-risk study? Why was the REC content for the first applicant's revisions to be reviewed and signed off by the Chair within the 60-day time limit, while those of the second applicant had to be scrutinised by the whole committee, requiring that the application be rejected, the regulatory clock stopped, and the revised PIS submitted anew as part of a fresh application? The key difference lies not in the risk of the interventions being proposed but in the two (both male) applicants' performance when they appeared before the REC. The first applicant was blandly neutral, accepting the REC's authority, answering questions in a straightforward manner and addressing an omission from the PIS in an appropriately apologetic manner. Thus, despite the extensive concerns about his PIS – seven different members contribute to the initial discussion of its weaknesses – the committee trust him enough to take the revisions seriously, only requiring the changes to be signed off by the Chair. The second applicant adopts a very different tone. Rather than simply acknowledging the complexity of the study, he seeks to dismiss the REC's concerns: 'we've never had a recruitment problem with similar studies'. Instead of acknowledging the significant amount of blood being sampled for testing, he quibbles over whether this matters because it was made up of a series of smaller samples. In both of these instances the applicant may well have been factually correct – his team may well have successfully recruited for similarly complex studies in the past, and the blood may well have been taken over time, in numerous small samples – but he was morally tone-deaf. The nature of his responses gives the impression of someone unwilling to take the REC process seriously, someone not to be trusted, an impression compounded by the offhand way in which he dealt with the issue of partial payment to volunteers.

Just because many REC members value the attendance of applicants at meetings does not mean, of course, that there is not criticism of this aspect of the review process. At one extreme is Rob, an expert member of Coastal MREC, who is sceptical about the value of applicants' attendance before RECs, questioning the subjective nature of the committee's evaluations of the interactions and its lack of formal expertise in assessing applicants in an oral setting. For this member, these problems can be found in the way in which an applicant who seems a bit arrogant or badly briefed can damage their prospects. This emphasises the potentially subjective elements of a REC's trust decision, and the inherent problems resulting from attempts to interpret people's behaviour as signs of underlying trustworthiness; an applicant's nervousness may be interpreted as a sign of evasiveness. For Rob the key issue is: 'What have we got here if somebody gives us assurances orally? What does that mean? Nothing.'[17] While other members might value attendance more than this member, they also acknowledge the problems:

> how much the actual person influences what they're saying as well, the attitude of the person, the way they're presenting things. It's almost like having an interview where you may not show the best of your ability because you're giving a poor interview in spite of the fact you're a very good candidate so it is a dangerous thing as well as a useful thing (Roderick, St Swithin's).

Similarly, Caroline, the administrator at Coastal, expresses

> concern about having researchers in sometimes ... they [committee members] judge what they see as the inadequacies of the researchers, and it sometimes borders on ... not being personal, but it's just like you're assessing that person's personality for five minutes, and their capabilities for actually conducting the research ... and they can turn around and they can say, well, he wasn't very impressive, was he? And if he's like that with us, how's he going to do this?

At the very least, appearing before a committee is seen as a potentially intimidating experience for an applicant, with Martin, Chair of Northmoor LREC, admitting to initially being 'very dubious' about inviting applicants to attend and acknowledging that at the start

> we had difficulties of some members actually slightly intimidating the researchers who were, who were coming up, particularly student researchers, so because of that, I made a rule in our committee that I would always gather opinions before, and that I would always do questions myself, so that I was in control of that ... I think it's a difficult experience to, to be, have questions fired at you by a group of nameless individuals around the table.[18]

However, these concerns are about judging applicants too harshly on the basis of attendance; there appears to be little consideration of the possibility that applicants might 'over-perform' their trustworthiness. Yet, as we might expect from previous scholars' work, the nature of organisational failure means that those features of REC review that help members make decisions they see as justified will also, occasionally, lead to disasters.[19] Unsurprisingly, members' accounts of the decision by the Brent REC to approve the TGN1412 trial underline the considerable part played by the applicant's performance of trustworthiness. The first aspect of the applicant's performance was that, unusually, the PI for this project (who worked for the contract research organisation (CRO) Parexel, and who the REC knew well from previous applications) was accompanied by a representative of TeGenero, the pharmaceutical start-up that had developed the drug. For members of the REC, this was a significant trust warrant, indicating that the researchers were taking the ethics review process seriously. As one member of the committee put it to me, 'We think it is nice if the sponsor comes along, because it just shows that they're that little bit more engaged as well.' But from the perspective of the applicant, the presence of the TeGenero representative was less a trust warrant indicating ethical engagement on the part of the sponsor company, and more an indicator of the PI's lack of confidence about his own knowledge. As one of his colleagues told me:

> And on this occasion, because he [i.e. the PI] didn't have the full confidence that he was going to be able to answer their scientific questions – he's not an immunologist – he asked the sponsor to supply somebody to come along with him. He said, 'I can answer their questions about this practicality and that practicality, but when they ask a question about *is this going to have a major effect on interferon one* and *tell us exactly why you think this drug is not going to be a dangerous thing*, I'll hand over to my colleague.'

In terms of the TeGenero representative's actual performance in front of the REC, one member told me:

> So he came in and the chap was very, very charming, very eloquent and knew all his data. And we asked him specific questions about was this relevant to, to humans … I think [member name] specifically asked him about the, what's it called, the storm…

AH: Cytokine storm?

> Cytokine storm, and he answered the questions very, very reassuringly … And I tend to come from a quite critical point of view to kind of raise potential problems and so I do sometimes kind of worry within the group dynamics that I'm kind of causing too much

trouble when I bring things up ... he gave a fantastic performance, there's no doubt about that, he was, reassured everybody and he reassured me.

For this member, a key feature of the representative's 'fantastic' performance of competence and trustworthiness was the belief, bolstered by his performance, that he was a physician:

> and I have to say I kind of assumed he was a physician researcher. And part of that assumption is because we're used to dealing with the CROs ... and they are clinicians as well as running it. So it wasn't something that we, I, certainly, for me, didn't think to wonder about. And the way he spoke and this kind of understanding of physiology, I just assumed he must be a doctor. And it was all, he seemed very confident that it was all going to be fine and that the animal data was exceptionally reassuring.

This assumption about the medical background of the sponsor's representative – and the associated assumptions about professional standards towards patients – was, however, misplaced. The individual in question, despite his persuasive performance, was a research immunologist with no clinical experience, rather than a doctor. In this case, the applicants' performance served to mislead the REC into a false sense of security regarding this specific application.

This account of the persuasive role that meeting the PI and company representative played in the Brent REC's decision to approve the TGN1412 trial underlines the broader value of applicant attendance in REC decision making; in essence the importance of meeting researchers face to face – 'getting to see the whites of their eyes' – in deciding whether to trust them or not. While REC members' reflection on applicant attendance tends to highlight both the value of meeting researchers and the possibly unfair nature of interpreting trustworthiness from personal performance, there is a general lack of discussion regarding the potential 'overperformance' of trustworthiness.

More broadly, the insight this chapter offers as to the importance of interaction and interpersonal trust is only partly compatible with typical discussions about trust in modern social systems, where the focus tends to be on the impersonal and systemic. Even Anthony Giddens' concept of 'facework' is not standalone but 'primarily a translation mechanism for trust ... link[ing] individuals to wider social systems, ... transform[ing] trust in an individual into trust in the system'.[20] Yet, in the case of RECs, such translation is rare – trust in an individual researcher rarely converts into broader trust in an applicant's department (or institution). Even the case in Chapter 2, where Coastal MREC discusses an application in terms of being from a 'good stable', the committee's trust is clearly invested in the specific applicant – Professor Gibbons – rather than a faceless institution. In the context of REC meetings, 'facework' is centred on those who attend the meeting, it is a personal judgement and tells us little about broader trust (on the part of the committee) in specific research institutions, hospitals, or universities. Crucially, this emphasis on the importance of interpersonal trust is not a residual feature of 'traditional' REC practice – although a longstanding aspect for a minority of committees – but rather a more recent feature, developed as a result of the modernisation and centralisation of REC activities, and in part because of the influence of pharmaceutical companies. Applicant attendance became a standard part of REC practice as a result, in large part, of requests by pharmaceutical companies, and has been enhanced by a variety of changes arising out of the steady centralisation of the REC system, such as allowing applicants to choose the REC they wish to apply to, or limiting the numbers of applications per meeting, thus giving time for attendance. Given ethics review's status as something of a regulatory anomaly, with strong roots in professional self-regulation, it is perhaps unsurprising that such a defiantly pre-modern assessment tool – interpersonal contact – would play such a large part. Yet, given these procedural changes' origins in very modern regulatory motivations – efficiency of communication, speed of decision, depth of reviewing resource – perhaps REC reviewing needs to be seen as more typical of how modern expert systems are actually regulated.

The previous three chapters have mapped out the different kinds of knowledge that RECs draw on – written, local, and interactional – to help them make decisions about the trustworthiness of applicants. The next chapter takes a slightly different approach, aiming to explore a longstanding area of disagreement around REC practice – the scientific review of applications – and the role of trust decisions in its insoluble nature.

Notes

1 Robert Levine, 'Inconsistency and IRBs: Flaws in the Goldman-Katz study', *IRB: Ethics & Human Research,* January/February 6:1 (1984), 4–8.

2 Steven Shapin, *The Scientific Life: A Moral History of a Late Modern Vocation* (Chicago, IL: Chicago University Press, 2008), 302.

3 These varying views of applicant attendance were also held by Sarah Dyer's interviewees with the 'majority of committee members ... interviewed [feeling] that the main advantage of researchers attending was primarily procedural. It enables a quick clarification of any questions the committee has' while at the same a couple of them were also feeling 'very strongly that it was important to see researchers especially where they were unsure about what they saw on paper. They described this in terms of, for example, the need to look the researcher "in the whites of their eyes" (C12) in order to judge their integrity and "see if we trust them" (C23)'. Sarah Dyer, 'Applying Bioethics: Local Research Ethics Committees and their Ethical Regulation of Medical Research' (PhD thesis, University of London, 2005), 95 and 96.

4 Laura Stark, 'Reading trust between the lines: "Housekeeping work" and inequality in human-subjects review', *Cambridge Quarterly of Healthcare Ethics,* 22:4 (2013), 396. Stark's review suggests that the problem is *not lack of invitations* from IRBs since, when busy committees invited researchers along, 'the invitations were rarely accepted'.

5 NHS Executive West Midlands Regional Office, *Report of a Review of the Research Framework in North Staffordshire Hospital NHS Trust* (Griffiths report) (Leeds: NHS Executive, 2000), 40.

6 What data there is suggests that attendance at REC meetings improved the chances of one's application gaining approval: Peter Heasman, Philip Preshaw, and Janine Gray, 'Does researchers' attendance at meetings affect the initial opinions of research ethics committees?', *Research Ethics Review,* 4:2 (2008), 56–58; Peter Heasman, Philip Preshaw, and Chris Turnock, 'Does attendance of students and supervisors at meetings affect the opinions of NHS research ethics committees of student projects?', *Research Ethics Review,* 5:3(2009), 101–103. Attendance at US IRB meetings – however rare – produces efficiency gains, but not necessarily improved approval rates: Holly Taylor, Peter Currie, and Nancy Kass, 'A study to evaluate the effect of investigator attendance on the efficiency of IRB review', *IRB: Ethics & Human Research,* 30:1(2008), 1–5.

7 More explicitly, one of Sarah Dyer's interviewees notes the insight meeting an applicant can give into their 'sloppiness of thought. And the way they dismiss things. We have had situations where we have been interviewing somebody and they come out with a way casual comment: "Oh, it doesn't matter how many blood samples you take". Well, it does matter, it matters very much'. Sarah Dyer, 'Applying Bioethics', 97.

8 For an exploration of the role of formal questioning in testing trust-worthiness, see: Charles Bosk, *Forgive and Remember: Managing Medical Failure* (Chicago, IL: University of Chicago Press, 2003, second edition), 94–102.

9 Anthony Giddens, *The Consequences of Modernity* (Stanford, CA. Stanford University Press, 1990), 80.

10 Amel Alghrani, Paula Case, and John Fanning, 'Editorial: The Mental Capacity Act 2005 – ten years on', *Medical Law Review*, 24:3 (2016), 311–317; Mary Dixon-Woods and Emma Angell, 'Research involving adults who lack capacity: How have research ethics committees interpreted the requirements?', *Journal of Medical Ethics*, 35 (2009), 377–381.

11 Although, as Chapter 4 shows, RECs are often quite willing to challenge, or at least comment on, the science of a proposal.

12 A possible issue here is that the speaker has a strong accent from elsewhere in Europe and, while he is very articulate, it is clear that English is not his first language. So the slightly off-hand manner of this applicant's presentation may be a function of his status as a non-native English speaker, rather than any reflection of his trustworthiness with regard to this study. Yet the REC has little or no way of knowing that.

13 Sarah Dyer, 'Applying Bioethics', 98.

14 In the case of surgery, for example, see: Joan Cassell, *The Woman in the Surgeon's Body* (Cambridge, MA: Harvard University Press, 1998).

15 In considering the potentially gendered aspects of performance, this example refers to two men, and we might wonder whether the language of the critique that Charlotte offers here – of 'cockiness' and 'arrogance' – would be applied to female applicants.

16 Good examples of the 'asocial' concept of risk in ethics review can be found in Annette Rid, Ezekiel Emanuel, and David Wendler, 'Evaluating the risks of clinical research', *Journal of the American Medical Association*, 304:13 (2010), 1472–1479; Julian Savulescu, 'Two deaths and two lessons: Is it time to review the structure and function of research ethics committees?', *Journal of Medical Ethics*, 28:1 (2002), 1–2. From a US perspective, good examples of this kind of reasoning can be found in: Charles Weijer, 'The ethical analysis of risk', *Journal of Law, Medicine & Ethics* 28 (2000), 344–361.
    This understanding of risk also underpins many social scientists' complaints about REC review: Jennifer Burr and Paul Reynolds, 'The wrong paradigm? Social research and the predicates of ethical scrutiny', *Research Ethics Review*, 6:4 (2010), 128–133; Elizabeth Murphy and Robert Dingwall, 'Informed consent, anticipatory regulation and

ethnographic practice', *Social Science and Medicine*, 65:11 (2007), 2223–2234; Helen Minnis, 'Ethics review in research: Ethics committees are risk averse', *BMJ*, 328:7441 (2004), 710–711.

17 Similarly trenchant views can be found amongst Sarah Dyer's interviewees: 'Some members felt very strongly that it was no business of the committee to be making judgements about the integrity of the researchers. They felt the committee was there to ensure that the proper procedure had been followed.' Sarah Dyer, 'Applying Bioethics', 96.

18 For a brief account of how intimidating appearing before a REC can be, see: Elizabeth Fistein and Sally Quilligan, 'In the lion's den? Experiences of interaction with research ethics committees, *Journal of Medical Ethics*, 38 (2012), 224–227.

19 The best review of this literature can be found in: Diane Vaughan, 'The dark side of organizations: Mistake, misconduct, and disaster', *Annual Review of Sociology*, 25 (1999), 271–305.

20 Frens Kroeger, 'Facework: Creating trust in systems, institutions and organisations', *Cambridge Journal of Economics*, 41 (2017), 491.

4

Reviewing science, trusting the reviewers

'If the experts do not agree with the peer review, the committee must
decide whether to take the existing review on trust.'

Sarah Edwards[1]

'and there's this constant debate going on … how much is the role of
the Ethics Committee to actually check your science and I suppose
I come down on the side of, to me, science has got to be OK otherwise it's a waste of everybody's time.'

Adrian (expert, St Swithin's Local Research
Ethics Committee (LREC))

The members of St Swithin's LREC turn to the next application, a
study comparing the success rates of two forms of surgery for a particular cancer. For the committee this raises concerns about the risk
of 'seeding' other parts of the body with cancer cells. Among the
committee members there is a vigorous debate about the proposed
research and, when the applicants attend, an extensive discussion
takes place, mainly revolving around statistics and the randomisation of the groups, as well as confusions produced by the applicants' apparent difficulties in filling in the form.

Following the applicants' departure from the room, and sparked
by the statistical issues raised by the proposal, especially around
the randomisation of the groups involved, there follows a discussion within the REC over the need for scientific peer review of REC
applications. Ingrid, an expert member, asks the Chair, Howard,
about something called the 'Warner report', and how it would impact on what was being planned over RECs' right to consider the
science of applications. She points out that the REC needs to know
that peer review has taken place, and that ensuring the independence of such peer review was hard. Howard replies that RECs have
'a long way to go before we abrogate responsibility on science' to
external peer review, giving the hypothetical example of a five-year
programme grant which might get reviewed but with individual

projects within the grant, perhaps taking place in year three, described by only a single line in the proposal, technically counting as being peer-reviewed. It is clear that, under these circumstances, it would be irresponsible of a REC to consider such a project as adequately peer-reviewed. The aforementioned 'Warner report' – or to give it its correct title, the *Report of the Ad Hoc Advisory Group on the Operation of NHS Research Ethics Committees* – arose out of the review of REC practice by a committee set up by Lord Norman Warner, a Minister of State in the Department of Health. References to it pepper my fieldnotes, with one expert's discussion of a proposal's inadequate definition of pancreatitis being prefaced with the acerbic aside: 'The sort of thing I now want to mention will be removed if the Warner report is enacted.' More expansively, Martin, Chair of Northmoor LREC, confided that:

> there is a feeling amongst RECs that the Warner [report] was heavily influenced by drug companies, and lent on by drug companies, and there were no representatives of RECs on the committee that made those decisions, so we, we don't really feel we've got any kind of ownership of that decision and it very much feels like senior people from above saying 'lay off the science'.

Yet, despite the crystallisation of concern around the 2005 *Report*, these debates cut to what is, at four and a half decades, one of the longest-lasting and most controversial debates over REC decision making, and the place where views of REC members are most clearly in opposition to those of policy makers and critics: whether RECs should take the quality of a proposal's science into account when deciding whether it is ethical or not.

The aim of this chapter is twofold. The first half explores the historical and intellectual roots of RECs' longstanding insistence that they should take the quality of science into account when reviewing the ethics of proposed research. This is followed by an exploration of the resulting puzzle: if science is so important, why then are RECs so reluctant to draw on external scientific expertise which might help them assess applications, with the answer lying in the importance of trust decisions to RECs.

To start thinking about the role of scientific review in ethics, let's return to the Ad Hoc Advisory Group, set up in November 2004 with the remit to 'provide independent advice to … Ministers on

the operation of NHS [National Health Service] Research Ethics
Committees (REC) ... [and] ... take stock of developments and
trends affecting the remit, administration, operation and workload
of NHS RECs in England.' The group's terms of reference paid
particular attention to 'regulatory blocks impeding research' with
a need to consider how to 'reduce the time required of researchers
starting high quality research' and 'strengthen the systems, structures and processes supporting NHS RECs to make their business
process as efficient as possible and improve users' and committee
members' experience of it'.[2]

Given that the UK ethics review system had just undergone a
comprehensive revision following the previous year's implementation of the European Clinical Trials Directive, it might seem an
odd time to commission such a review. Indeed, it was discussion
of the Directive in Parliament that triggered the review; in a series
of debates in the House of Lords, a number of objections were
raised about the way in which the government had implemented
the Directive. While some of these claims, for example, that the
centralisation of the system involved undue political influence over
RECs, could be ignored, concerns about pressure on ethics review
and subsequent impact on medical research were harder to dismiss.
Claims that 'some RECs have found themselves under pressure',
that lack of resources might lead to 'an absence of consistency'
in REC decision making and that 'the requirements laid down by
some research ethics committees are becoming ever more burdensome and time consuming' persuaded the Minister to seek independent advice.[3]

While the resulting *Report* covers a range of issues relating to
ethics committees, it was its recommendations around scientific review that were the focus of REC members' concerns. The Warner
committee noted that:

> RECs are not, and should not be, responsible for detailed scientific
> review ... Because it is unethical to conduct scientifically inadequate
> research, the RECs' role is to be reassured that there has been ad-
> equate scientific review of the design. For most applications, this
> review will have taken place before the application reaches the REC.
> We do not believe RECs should function as a secondary form of sci-
> entific review; indeed, to do so would have significant implications
> for REC membership.[4]

Yet, with regard to scientific review, the Warner report did not introduce any new elements to discussions around the roles and function of RECs; it simply restated a longstanding position held by the Department of Health and its predecessors. Indeed, one can trace tensions around REC members' ability to review the science in an application back to the foundation of RECs in the NHS in the late 1960s.

While the issue of scientific expertise was used by the Ministry of Health in its early attempts to exclude lay members from RECs, the role of scientific understanding in the assessment of ethics applications became incorporated into discussions about how these groups should make their decisions.[5] Until the early 1990s, when the Department of Health finally accepted formal responsibility for RECs, the main authority offering guidance on REC practice, composition, and decision making was the Royal College of Physicians (RCP). Following its initial engagement with issues of ethics review there was a gap of 11 years before the College formally returned to the topic of guidance for RECs, during which concerns arose that 'some Ethics Committees did not know what they were doing' and there was thus 'a need for general guidance on how Ethics Committees should be run'.[6] The resulting 1984 report, the *Guidelines on the Practice of Ethics Committees in Medical Research*, was the first attempt to comprehensively set out the boundaries of REC composition, decision making, and authority in the NHS.

During drafting, comments were made by the Fellows of the College (at meetings known as 'Comitia') with regard to scientific review, noting that: 'it was a very good document but a little weak-kneed about the scientific quality of projects to be undertaken. It was unethical to do something that was badly or inadequately designed.' Desmond Laurence, who had taken the lead on drafting the *Guidelines*, agreed: 'it was a little weak-kneed but this was a very controversial area'. Expecting RECs to send applications out to external review when they felt they lacked the relevant expertise led Dr Laurence to 'shudder slightly at the thought of large numbers of investigations going to outside advisors'. Having requested a suitable form of words, Dr Laurence accepted an amendment to his report that stated: 'It would be open to Ethical Committees to disapprove a project if it were thought the scientific aspects were inadequately dealt with', noting that 'the message in the document was bad science constituted bad ethics. Such an amendment would strengthen Ethics Committees and make them see clearly that the science was their concern.'[7] Thus the RCP's 1984 guidelines emphasised that 'A committee should feel free to refuse an application on grounds of inadequate scientific quality.' While it admitted that 'the question of the extent to which scientific quality, design and conduct should be considered continues to cause difficulty', in the end it concluded that 'badly planned, poorly designed research that causes inconvenience to subjects and may carry risks without producing useful or valid results, is unethical'. While

> full scientific evaluation is beyond the capacity of the great majority of Ethics Committees ... Nonetheless, Committees should approve only studies which are of good quality ... do[ing] their best in this area, using their own knowledge and knowledge of their colleagues.[8]

Subsequent guidance from the Royal College, published in 1990 and 1996, while acknowledging the limitations in RECs' expertise, reiterates the need for questions of scientific quality to be taken into account in ethical review.[9]

As we saw in Chapter 2, by the late 1980s, as a result of a number of factors, the Department of Health was coming under pressure to introduce a degree of standardisation into the REC system. In 1991, when the Department took formal responsibility for RECs for the first time, the acceptance of RECs' need to perform some sort of review of the scientific content of an application subtly shifted. RECs now needed simply to know that an application had been adequately reviewed by someone, asking themselves 'Has the scientific merit of the proposal been properly assessed?'[10] Subsequent guidance from the Department of Health underlined this shift, stressing that the role of a REC was not to review the science of an application, but rather to ensure that adequate scientific review had taken place. For example, the *Governance Arrangements for NHS Research Ethics Committees* (*GAfREC*) document, which set out the general framework for RECs from 2001 onwards makes clear that:

> It is not the task of a REC to undertake additional scientific review, nor is it constituted to do so, but it should satisfy itself that the

review already undertaken is adequate for the nature of the proposal under consideration.[11]

Far from the 2005 Warner report introducing an unusual and novel requirement into REC practice, it simply restated a well-acknowledged feature of official thinking about RECs.

Yet for many commentators, this separation of ethics from science is barely credible. After all,

> science and ethics are not mutually exclusive, so that design considerations such as the exclusion of minority ethnic groups from research, the inclusions of a placebo control group when standard therapy is available and the availability after the trial ends of a treatment found to be effective have ethical dimensions.[12]

As one commentator points out, when thinking about the risks raised by proposed research, ethics review bodies often end up thinking about the quality of the science:

> in some cases what prompts a question about scientific benefit is the nature of the risk: if the risk is unmistakably unreasonable (unnecessarily nonminimal, avoidable, or far out of proportion to purported benefits), the IRB members are likely to express doubts about the quality of thinking that went into the science. If the assessment of risk cannot be neatly isolated from the assessment of scientific benefit, then a strict comparison or a meaningful risk–benefit ratio is much more difficult.[13]

Echoing these arguments, it is clear that, for members of these committees, as one interviewee put it: 'You can't separate the ethics from the science.' Roderick, an expert member of St Swithin's LREC, approached applications 'from the point of view that if the science isn't any good then you are wasting the patient's time … and therefore not being ethical in your treatment of the problem', a point reiterated by Rob (expert) from Coastal MREC, who suggested that:

> If you've got a bad scientific project and particularly if you've got a project that is statistically so poorly designed that it will not get its reported effect, then it is ipso facto unethical because you've wasted people's time and money.

Given that researcher scientists make up the largest single group on these committees – 37% of members are clinical researchers, with a further 28% conducting some kind of research on human subjects – and 'Indeed they are on the committees because they are research scientists ... It is ... not surprising they should have "firm opinions" about good research and should include these in their assessment of applications.'[14] As Celia (expert, St Swithin's) suggested,

> If there's nobody there that's got any idea about the science at all, because if you want to separate the science entirely, why bother to have any scientific people on an ethics committee? Why don't you just have, you know, members of the public?

Thus RECs often ignore the idea that they should not regard scientific quality as the basis of their decisions, with one study noting that scientific issues were raised in 74% of reviews. More specifically, these concerns cropped up in all of the sampled applications that were initially rejected and 96% of the applications that were initially given provisional status but then finally rejected, 'suggesting that issues of scientific quality were strongly associated with applications that were initially or eventually deemed unfavourable'.[15]

For many members, a key aspect of the science concerns the statistics applicants claim will be used to analyse data produced from research and the way in which a research project is designed in order to produce that data. For example, while Northmoor LREC is discussing whether to reject an application or not, Damien, an expert member, suggests that he 'would definitely encourage [the applicants] to resubmit because the results could be interesting, but it comes back to the science thing'. Amanda, another expert, suggests that the REC could delay a decision by 'wait[ing] for a statistical report'. Charlotte, a lay member, tries to steer the committee away from a reject – 'the ethical issues in this study are less than the last study which we passed' – but Megan (expert) disagrees: 'It *is* an ethical issue to do research on people that won't tell you anything.' At this point Damien tries to open up the discussion further – 'But it's not just the stats. It's that this [study] population will not match the disease group' – only for Martin, the Chair, to bring the debate back to the issue of statistics 'I can't myself see an ethical problem with this, apart from the statistics.' The committee finally decides on a rejection. The debate moves back and forth, and, perhaps unsurprisingly given the wider debates around the role of scientific review in ethics decisions, some members question whether the statistical weaknesses are important. Yet despite this range of views being articulated, there is core concern about the ethics associated with the statistics presented in this application.

While obviously there are issues around *over-recruiting* for research, this is rarely seen as a problem by RECs.[16] More commonly the issue around power calculations centres on whether the sample size is large enough. As Gretchen, the statistician for Coastal MREC, bluntly pointed out in discussion of a particular trial, 'in terms of statistics, 300 people is far too small for the ambitious claims being made for the study'. In another example, the MREC highlights how issues around study size constrain the design of a project (in this case, specifically the nature of a successful outcome for the study) and in turn bleed into problems around the possible risk–benefit trade-off of this research: Colin, the lead reviewer, turns to Gretchen to ask 'whether the statistical question is still a riddle or does it make some sort of sense?' She seems equivocal, pointing out that there is also a problem with the study's design, since the application does not define what counts as a successful outcome: and as a pilot study – given the small numbers involved, that's what is being proposed – this is crucial. When Colin suggests that there is something problematic about putting people who may be vulnerable through two diagnostic tests that may be of no value, Gretchen responds:

> and one of the tests could be harmful. They certainly haven't explained what analysis they are going to do. They simply say: 'some statistical methods'. If you think that the stress test is going to be harmful then you may be concerned as to whether to do this.

That such statistical limitations should play a key role in REC decision making is not surprising given that statistics is a key expertise that RECs are encouraged to have. The Department of Health allows that a statistician is one of those 'member[s] provid[ing] unique expertise to the REC' such that committees may appoint a 'deputy' member to sit in and provide expertise in those meetings at which the main statistician cannot attend.[17] In terms of the interactional aspects of REC decision making, the importance of statistics translates into considerable personal authority for the statistician themselves. As one member of Coastal MREC noted, while RECs have avoided relying too much on specific kinds of expertise (what he refers to as 'specialist dependence'), statistics is the exception: 'we're not able really to do anything unless the statisticians said something'. So, although the committee itself is 'not that statistically illiterate', it remains the case that 'I think that we really are now in the position where the only indispensable member is the statistician' (Rob, expert). In the RECs I observed, it is not clear that this indispensability translates into some kind of personal power – it is possible for committees to approve proposals despite statistical concerns – but it is reasonable to view the statisticians as having particular authority.

And perhaps underpinning this authority is the fact that, in the case of clinical drugs trials, consideration of the design of the trial by the REC is legally required.[18] Thus statistics is a clear point of tension, not just between the views of the RCP and the Department of Health, but within the Department of Health's own rules: at the same time as there is a legal obligation to review the design and statistics of a pharmaceutical trial, RECs are discouraged by the Department of Health's guidelines from carrying this kind of scientific review.

Yet beyond statistical issues being closely integrated with risk–benefit decisions, the very nature of REC decision making means that scientific issues and ethical issues are difficult to separate in any meaningful sense. As one expert member of Coastal MREC put it:

> I look at all of the protocols from a scientific point of view because I think that the science and the ethics are intertwined. ... a scientifically flawed study can't be considered to be ethical, so I don't agree that they should be separated. I think it's very difficult to actually separate out the science from the ethics (Gretchen, expert).

The entanglement of science and ethics immerses trust decisions in a technical appearance. As might be expected from Chapter 3, how an applicant responds to queries about scientific quality can serve as a trust marker. For example, in one particular discussion in Coastal MREC Daniel asks: 'If we get an unsatisfactory answer on the objective measure, what do we say – is it an approve?' Rob, the expert member who took the lead on this application, responds: 'I think it's an important scientific handicap but not an ethical handicap', prompting Adam to point out that 'Part of the answer depends on what they say. If they've thought about it then it's OK, but if they just stick two fingers up then I would get suspicious.'

In part, concern about the scientific quality of an application is a way of exploring issues around the crucial issue of clarity, echoing issues explored in Chapter 1:

> In my experience we drift into the science when we are not clear about what the benefit is and therefore whether the cost to the individual is worth the benefit. And that's when we then, rather than saying 'no, it's not clear', we try to unpick what we understand from the application about the science.

This interpretation, from Louise, an expert member of St Swithin's LREC, is supported by Rachel, an expert from Coastal MREC:

> I think the other set of issues is put under this umbrella about the science versus the ethics ... perhaps some are mislabelled as being about science, and are actually about clarity. And I think those aren't about a specialist taking a view on the minutiae of a particular area of clinical specialism. I think they're actually about a more generic researcher saying, 'well, actually, it isn't clear how this thing has been designed'. ... if that isn't clear, then you can't really picture quite how the research is going to be conducted and, therefore, I think that gets in the way of fully getting a sense of the ethics.

Yet while sometimes a committee's concern about science is a way of working through a particularly unclear application, it also is the case that poorly designed research is, in itself, seen as ethically problematic. This would be the case in those examples 'concerned with the direct effects of poor study design on the well-being of people, such as the possibility that people in the placebo arm of a trial would be denied a known effective treatment ... for no clear benefit'.[19] Returning to a case discussed in Chapter 3, take the application to Coastal MREC for a trial of a compound to treat cardiovascular disease, with the drug being tested administered alongside either a placebo or one of two different doses of a particular statin. The application proposed recruiting patients, some of whom would already be stabilised on a statin (not necessarily the one in the trial). Simon, a clinician and expert member of the committee, expressed his concern that people get the best treatment – people may be taken off one statin, on to another statin at a lower dose. This needs to be mentioned in the PIS, even if such a change is not a problem for the short time the study is planned for.

When the applicants attend the meeting, this point is brought up by another expert member, Sarah, who tries to examine the scientific rationale of using this particular drug and the use of different doses. The point is that a patient normally on a high dose of one statin could be moved on to a low dose of this, or another, statin for the study, and a patient normally on a low dose could be moved on to a higher dose, in each case, possibly of a statin they are not stable on. The former run the risk of reduced efficacy, the latter of side effects. Although, to a large extent, this is a scientific discussion about the correct design of a trial, Sarah phrases it in terms of patient protection, about side effects: 'the point is risk/benefit'. As previously noted, the applicant's performance in this case was problematic – the REC remained unconvinced and chose to reject this application.

This debate clearly touches on the technical, scientifically based design of a clinical trial. It is a discussion about the science. Yet it is also a discussion about the ethics of this trial, about whether the design in question involves too great a risk for the potential benefits. Even authors critical of cases where RECs' scientific comments can be deemed trivial or unnecessary suggest that cases such these 'legitimate the interest of RECs in scientific issues and strengthen the argument that RECs should adjudicate on these matters'.[20] And those more sympathetic to REC review of science, such as Richard Barke, point out the unavoidable

> intertwining of ethics and research design [that] is illustrated by the concept of external validity. The generalizability of research findings to larger populations is a fundamental issue in research design: it depends on the size and selection of the study sample, but also vice versa. A research design embodies decisions about the intended impact of the study, and the inclusion or exclusion of particular subpopulations is inherently an ethical issue.[21]

Of course members are aware of the controversial nature of such debates and occasionally act to 'police' their colleagues who may be seen as going too far in the direction of a scientific review. When

Neil, an expert on St Swithin's LREC, suggests that, in the case of the application under discussion there are problems with the design of the research (a study into treatment for hip mobility problems), that 'the one or two hip issue is a confounding factor', Charlotte (lay) argues that his comment crosses the science/ethics divide: 'It's not our role to comment on this. If the science is skewed it will be picked up when they try and publish in a journal.' Damien chips in: 'But part of our role is that the science has to be good', only for Charlotte to reply: 'Or good enough?' Amanda, another expert, weighs in: 'I wouldn't want to reject it on those grounds.' A similar, if more heated, discussion takes place in Coastal when Rachel, an expert member, interrupts Pete, who is acting as lead reviewer on an application, stating that she is very concerned that the REC is going into detail on the science of the study, which is funded by a well-respected research charity and which has four referees' reports included: 'I really feel that we're straying into territory that we as an ethics committee are not qualified to do.' Pete responds, noting that the point of this is to ask the researchers to clarify the factors in the imaging software they will use. 'I am not prepared to *not* comment on an application – I have a right to express these views, otherwise there's no point in my being here.'

While one perspective on this exchange is that this is a good example of a REC acting responsibly, restricting the comments it will make, it does raise the problem of previously conducted external peer review. And at this point in this chapter, our focus shifts from the challenges around whether RECs should perform scientific reviews, to more trust-based issues, exploring the idea that they should rely on external reviewers instead, as a way of solving this dilemma.

As already noted, the Department of Health's solution to the problem of scientific quality is to ask RECs not to perform a scientific review themselves, but rather, given that 'protocols submitted for ethical review should already have had prior critique by experts in the relevant research methodology', the role of a REC is to 'satisfy itself that the review already undertaken is adequate for the nature of the proposal under consideration'.[22] Where there is a discrepancy, it lies in the difference between the guidance offered to RECs in 2001 and the more bullish suggestions arising from the Warner report four years later. The *GAfREC* guidance document suggests that, 'If the committee is of the opinion that the prior scientific review ... is not adequate (including adequate statistical

analysis), it should require the applicant to re-submit the application having obtained further expert review.'[23] However, with its suggestion that 'Where peer review has taken place, the RECs should accept this in all but exceptional cases', the Warner report questions whether RECs should even, normally, make judgements on the quality of scientific review.[24]

The Department of Health's position on the value of scientific review could be seen as simplistic, ignoring, for example, concerns over the 'halo effect' that research funding might produce. As long ago as 1986, the RCP Committee on Ethical Issues in Medicine debated the role of external scientific review, admitting that, while RECs 'are also supposed to take into account the scientific validity and purpose of the work ... as a rule they are not fully competent to do this'. Despite this:

> it was felt that a problem might exist if a grant is funded without prior approval by an ethics committee as the committee might feel itself under some pressure to approve an application if it is already known to be funded.[25]

Concern about 'the strong pressure that would be exercised on local ethics committees if applicants went to them after receiving approval of a project from the MRC [Medical Research Council]' and how this 'would threaten the essential independence of the local committees' was the reason given by the MRC for requiring grant applications to undergo ethics review prior to submission.[26]

Added to this is the different purposes served by different kinds of scientific review. For example, in the US context:

> the IRB is likely to have a different standard for judging the quality of scientific methodology than a conventional peer review panel. As one IRB administrator put it, 'I don't trust peer review: they look at only the parts of the proposals that they know best, and they miss the big picture.' The task of most peer review panels is only to decide whether the research is worthy of funding, not whether it will be done.[27]

The practical problems faced by RECs when dealing with external peer review are highlighted by a 2010 meeting of the UK's National Research and Ethics Advisors' Panel (an advisory group including experienced REC members) which:

raised the question of what RECs should be doing with any scientific review that they receive. Should they simply be checking who had conducted the review and if, for example, it was an organisation such as the MRC then simply be content that adequate review had taken place or should they be assessing the review process against published standards? ... the guidance for reviewers is variable and so there was a need to set out the issues that any robust peer review should have covered so that RECs can properly assess the submitted review. If these criteria are then published then applicants to RECs would know that these questions will be asked by the REC which should in turn lead to the submission of scientific review that met these published standards.[28]

One obvious problem with the suggestion that a REC should accept external reviews as a matter of course is that the committee 'would look rather stupid if it gave a favourable opinion to a study with serious scientific flaws, and it is difficult to see how an REC [sic] would be "satisfied with assurances" when there are obvious flaws in the design'.[29] However this needs to be balanced against the practical challenges facing RECs as they attempt to make decisions about large numbers of complex cases. As Michael Pegg, Chair of the Royal Free LREC, noted when discussing his committee's decision to approve Andrew Wakefield's research into measles, mumps, and rubella (MMR) vaccination:

> We have had a review by Professor Epstein [the scientific reviewer] ... That is it. At that time, in a year we could do a couple of hundred applications. We did not question what researchers told us. We did not have time ... We get letters like that every day, signed by eminent people. We are not policemen. We just believed what people told us.

As he stressed later on, 'If he, [i.e. Epstein] a senior professor at the Royal Free, says these are clinically indicated [tests], I do not see we have any place questioning that.'[30]

As well as these tensions around what to do with such a review when it arrives is the limited number of REC applications that actually include scientific reviews: in 2009, only 2.4% of drugs trials included a separate scientific review, with other kinds of research fairing slightly better, at 4.16% of applications. These low rates were partly due to committees' reluctance to require applicants to submit reviews: 'RECs had requested scientific review where none had been submitted on only six occasions' in all cases for non-drug trial applications.[31] To some extent these low levels stem from REC members' uncertainty about committees' powers to require such reviews. As Adam, a lay member of Coastal MREC, put it: 'Now, there's a system there about independent review that doesn't work, because it's not compulsory, or doesn't seem to be. And what we can do about that isn't clear.' Even in cases where committees do ask for and receive peer review, the utility of these documents can remain frustrating. As Louise, an expert member of St Swithin's, explained:

> last meeting was very interesting because we strayed down another line because this new ... fairly new to us is this notion of external peer review and we'd got ourselves into a bit of a pickle about what we think the peer review is about ... [The application] ... had something that actually we didn't understand, so we sent it for peer review, but the peer review came back and said 'jolly good piece of work, really important area, I support research in this area'. Okay, which actually wasn't helpful because *we* thought it was probably a jolly important area and needed some research. We just weren't sure about what was presented to us, would [it] produce any answers and ... the peer reviewer didn't answer that question, but we didn't know that they'd been asked to and we weren't sure they'd ever be asked to ... and we spent ages talking about peer review and it was me that stopped it, because I suddenly said, 'I don't think it's about peer review, I think it's about we don't understand why he's doing it. And if we don't understand why he's doing it, how can we approve it?' ... I don't have any problem at all saying that if I don't understand it, I can't approve it and actually, that's not my problem.

In this context, trust plays a crucial role in the decision about whether a peer review is adequate or not, with a key concern centring on the independence of reviewers from applicants and hence the trustworthiness of the review. The practical challenges raised by the need for independent reviews were spelled out by one REC member, who noted that:

> if you say that any project must have been subject to an independent scientific review, actually, who is going to do that? You can just transfer a bit of the review system, the scientific bit as it happens, out for someone else and you do not know whether, how are you going to audit that, how are you going to police that, how are you going to be satisfied that the independent review is independent?

While issues around external review were not there in every application, the committees I observed regularly wrestled with these problems. For example, in one meeting a member of Coastal MREC notes that, when asked about whether the statistics of an industry trial have been independently reviewed, the applicants 'suggest that a [company name] statistician counts as independent ... They should just say "no, the application has not been seen by an independent statistician".' The following month, similar concerns are raised in the same committee, when Adam, who is leading on an application, notes that 'They [the applicants] tick the box that says that there's been an independent, external review but don't provide the actual review itself.' The REC suspects that the applicant is implying that different arms of the company can count as external reviewers. When the applicants attend the meeting, Adam raises the issue of the internal/external review. The applicant responds by agreeing that 'It certainly reads like it has had external review.'

'But has it?'
'No.'

Adam sums up: 'In the light of the previous discussion, if there *had been* an external review then some of those issues [around the statistics] would have been addressed. I'm a fan of such reviews.'

The dilemmas around the adequacy of external peer review are built into the organisation of NHS REC review:

> it is not at all clear how RECs should satisfy themselves that the application has undergone appropriate review, since the current application form requires applicants to state what kind of scientific review has been undertaken but not necessarily to include the reports with their application.[32]

The problems of this approach can be seen in an exchange between Howard (the Chair of St Swithin's LREC) and representatives from a pharmaceutical company: 'What would your position be on our seeing the peer review?' The company representative tries to deflect the focus of this question, pointing out that 'there are external consultants on our peer review panel ... and this is the standard protocol for these drugs so any peer review would suggest that this is the way to go'. Howard is not so easily put off and returns to the committee's point: 'How are you as an organisation going to be able to provide us with peer review?' The final response from the company – 'Adapt' – does not appear to inspire confidence.

As Sarah Edwards notes, for 'trust in the peer-review to be justified, it must be informed', with RECs looking at 'the position, expertise, track record, and reputation of the peer reviewer' and the reviewer themselves having to be 'an independent "peer" who is close enough to the subject matter to give a critical opinion without being so close to the investigator that his objectivity can be bought into question'.[33] In practice, of course, almost none of this information is available to RECs. The key problem around decisions about the adequacy of industry reviews, for example, is the lack of transparency; an applicant can claim that a review has been carried out but then not include a copy of the review or any information about the reviewer. The relationship between applicants and reviewers is, as far as the REC is concerned, opaque. As Howard points out:

> then companies come along and tell you that you can't see their peer review anyhow, and when you do get peer review from a company you don't know whether that peer reviewer has been paid £1,000 to write the peer review article and, to a certain extent, if they're not being paid £1,000 and why *aren't* they being paid £1,000 because everybody else is being paid £1,000.

Given the intense levels of commercial sensitivity that characterise the pharmaceutical industry, this institutional secrecy is unsurprising. Companies claim such protection is required to protect intellectual property and recoup the costs of innovation. Critics however point to the ways in which excessive secrecy prevents the publication of results contrary to companies' interests.[34] This culture of secrecy is reflected at the regulatory level in the UK where pharmaceutical regulation was, until 1997, governed by the 1911 Official Secrets Act and where the 1968 Medicines Act made it a criminal offence to reveal evidence submitted to, or the internal deliberations of, a drugs regulator, a restriction explicitly aimed at protecting the commercial interests of companies submitting applications for review and encouraging such companies to make their drugs available in the UK.[35] Given this lack of transparency, it is perhaps unsurprising that RECs fall back on their own judgement about the quality of the science, rather than push applicants into providing copies of peer review. Graham (Coastal MREC, expert) put it, 'When you've got these hidden peer reviews, and hidden toxicological reports, you've got to look into the science, even by seeking further scientific opinion yourself.'

One of the problems about such a lack of transparency is that it provides a space within which mistrust of an applicant can arise. In one application seen by St Swithin's LREC, the applicant has not turned up, and Howard begins to summarise the committee's position on this study: the REC wants scientific justification for this research, access to reviews, and, crucially, evidence of what the reviewers *were asked to review*, with Ingrid (expert) suggesting that reviewers might have seen a very different scientific proposal to the one presented to the committee. This final concern is only really viable in a regulatory space where RECs do not have access to full information about such reviews.

It would, however, be misleading to suggest that problems around transparency are limited to industry-sponsored research. As Celia, a member of St Swithin's LREC, points out, academic research can be equally opaque in terms of the relationships between applicants and those they choose to review their science:

we often have a letter attached saying ... 'I've looked at this application and it seems that it's a valuable contribution and this, that and the other', but do we see that many that are in detail? ... and of course sometimes the scientists on the committee say, 'oh yes, well, you know, he's a colleague of his, he knows him', so you're very unsure...

In the wake of the Northwick Park disaster, the St Swithin's LREC, at least, was sensitised to issues around external review of academic sponsored research. On one occasion, while commenting on an otherwise acceptable application from a researcher well known to the committee, Howard notes that, while the study seems to have undergone internal review, this is 'probably verging on the inadequate, we may have to mention in the letter we send applicants that people have to "up the ante" in terms of peer review'.

While RECs are, in theory, restricted to assessing the quality of reviews, actually *testing* or questioning this quality is discouraged by applicants. For example, Graham, a long-serving member of Coastal MREC, reminisced about his previous LREC, where:

we ran into trouble with, in Phase I studies, where we had a very good relationship with [pharmaceutical company name]. We ran into trouble with them, because we, if there was a first in man study, we would always seek, at their expense, an independent toxicological review. And, you know, I think recent events have proven [laughing] that this is a very wise thing to do.

Even if RECs do not commission an independent review, but simply seek to confirm the content of a scientific review, applicants can complain. For example, Rose, Northmoor's statistician, noted that:

I find sometimes ... people say they have consulted a statistician. Sometimes they put down the name ... and when I look to see what they have done [in the statistical review] it's very suspect. Now then, there could be two reasons for this, or three reasons. One is that they're telling lies, which I think can be the case. Secondly is that they've misunderstood what the statistician has said, which I think is probably even more the case, but there is the possible chance that the ... that particular statistician and I don't agree. And that can be a bit worrying, and I then want to find out a bit more about the situation.

She then went on to relate a case where an applicant used excerpts from a statistical review written by someone she knew. The inaccurate nature of the comments in the review confused Rose since she knew the reviewing statistician to be scrupulous ('it would be very, very precise figures'). She attempted to contact the reviewer to clarify what had been reviewed and the advice given to the applicant, but could not get in touch with the other statistician. 'And then the next thing I hear from [the REC administrator] he'd complained. The applicant complained that I'd actually contacted his statistician.'

Given the complex and uncertain nature of the role of scientific reviewing – whether carried out by the REC or external reviewers – it is unsurprising that the events at Northwick Park in 2006 reverberated throughout REC meetings and in interviews with members. A key feature of members' accounts about the TGN1412 trial centres on how such a disaster could just as easily have befallen one of the applications they reviewed:

There but for the grace of God go us. I mean, we could easily have a study where we are slightly outside our expertise and we make a stupid decision, or we fail to see the danger. I mean, I don't particularly think they made a stupid decision, I don't know, they may or they may not have. But on the information they probably made a reasonable, they may well have made a reasonable decision (Howard, St Swithin's).

In the wake of this trial, committees openly discussed Northwick Park and how it might shape their attitude towards transparency in peer review. For example, as part of such a discussion at the beginning of a meeting, debate within St Swithin's LREC moves on to the issue of peer review. Howard suggests that 'we really ought to be ready to send things back for peer review, if we feel at all nervous' while Abigail (lay) points out that 'this is something we have been after for ages: adequate signature from supervisors and peer reviewers'. Howard then informs the committee that:

> I have stoked things up and told the [Hospital] Trust that they need to sort out peer review: we may carry on in the current way for a couple of months, but that's it. What we're going to suggest is that the body [i.e. the hospital] takes responsibility with a cover sheet clearly stating the level of peer review and an explanation for *why* that level of peer review. After Northwick Park, we should say to companies that either we see peer review or we reserve the right to send it out to our own peer reviewers.[36]

In reality, the Brent REC's decision to approve the TGN1412 trial was not as straightforward as members of other committees seemed to think. Indeed, in its deliberations the committee paid particular attention to those aspects of the drug that were problematic, and which resulted in the cytokine storm. When the committee met on 23 January 2006 to discuss this application there was some concern on the part of several expert members who were uncertain about the safety of the trial.[37] While there was expertise in related areas, and thus awareness of the complexity of immunological reactions, there was no immunologist on the committee, and as a result, while the Investigator's Brochure was quite reassuring about the animal studies, the committee was unclear how closely animal immune responses related to human ones, and how applicable the 'safety' data from the preclinical studies was to humans. As one of these expert members of the Brent REC suggested to me, while normally he was confident in saying whether he was happy with a trial or not, in this case he could not give a clear 'yes' or 'no'.

For many commentators, the crucial differences between the immune systems of the test animals and humans which led to such differences in response were obvious prior to the trial taking place, and the failure to distinguish this was the key flaw in both the TGN1412 trial and regulatory scientific review.[38] Although the drugs regulator – the Medicines and Healthcare products Regulatory Agency's (MHRA's) – clinical and preclinical assessment of TGN1412 discusses the differences between human and rodent immune systems (hence explaining why rodent test data was unreliable), it missed the important differences between human and monkey immune systems, suggesting instead that: 'pharmacokinetic (toxikinetic) characteristics of TGN1412 as determined in cynomolgus monkeys are assumed to be most predictive for human PK [pharmacokinetics]'.[39] It was these differences that led to such an extreme reaction in humans, despite little or no immune response in monkeys.

When the representatives from the contract research organisation and sponsor company attended the meeting to answer questions, the committee, apparently uncertain about the science and safety of the drug, asked the applicants to supply additional scientific reviews of the trial. As discussed in Chapter 3, to some extent, the performance by the principal investigator and the representative from TeGenero calmed REC fears. Nonetheless: 'The Chair asked [applicant name redacted] to submit to the committee a report from the immunologist along with a short CV for the immunologist.' One of the applicants 'responded that a toxicology report had been submitted as reviewed by [name redacted]' but the REC insisted, asking them 'to access a specific clinical immunologist to review the toxicology report'. While the committee was reassured somewhat by comments made by the TeGenero representative – he claimed that 'from animal experiments it was indicated that the safety of the drugs [sic] could now be assessed in humans' – the decision letter sent by the committee on 31 January 2006 still asked for more reviews.[40]

TeGenero therefore commissioned three additional reviews from clinical immunologists which were sent with their amended application. These reviews discussed the potential risks of TGN1412 in terms similar to the trial documentation. The review written by the company's own immunologist noted that: 'The risk of inducing a clinically apparent cytokine release syndrome (CRS) by adminis-tration of TGN1412 in humans is considered to be low, although it cannot be completely excluded', while the author of one of the other reviews 'consider[ed] the pre-clinical safety of TGN1412 as sufficiently documented to justify the first-in-man application'. The final review described TGN1412 as 'a promising therapeutic anti-body candidate' and the 'trial design as justified to assess the safety, tolerability, PK and immunological effects of TGN1412 in hu-mans'. When these reviews directly addressed the issue of the simi-larity between monkey and human immune systems, they asserted the claim, since viewed as inaccurate, that human and monkey im-mune systems were identical in the relevant immunological sense.[41]

The issue now facing the Brent REC was whether these re-views were enough for the committee to regard the trial as safe to proceed. The nature of this decision was restricted by the rules governing the amount of time RECs can take to reach their final de-cision on an application. While the initial decision to ask for the re-views was the result of a committee discussion, because of the time limit and subsequent pressures introduced by the Clinical Trials Directive, the decision as to whether these reviews were sufficient rested with a single member of the committee. Before the Directive was introduced in May 2004, such a decision would probably have been taken by a sub-committee of the REC, but the time limit has had a number of effects on REC decision making, the key one in this case being that the people who decided that they were unsure about the safety of this trial did not take part in deciding whether the applicant's response was adequate. In order to stay within the time limit, RECs have tended to reduce the use of sub-committees to check applicants' response, and now have a committee Chair or Vice Chair carry this out.[42]

As already noted, one key challenge related to deciding whether an external review is adequate or not centres on the opaque finan-cial relationship between the companies that commission these re-views and those that carry them out. While one of the reviewers was clearly employed by TeGenero, the relationship between the com-pany and the other two reviewers – including how much they were paid – is not obviously included in the information supplied to the REC. Yet this case highlights how making decisions about whether to trust an external review goes further than questions about who paid for it. Although in the past, the Brent REC had sought out its own independent reviews of applications, this was 'usually not on a fundamental safety issue ... Normally it's on some point in the protocol'; thus one of the problems with an external scientific review that adjudicates on a trial's safety is that this 'outsources' away from the REC a judgement which underpins the risk–benefit decision, a key aspect of ethical approval. As another member of the Brent REC put it, given poor scientific explanation in an application:

> we didn't feel that it was appropriate for us to defer to someone [i.e. a referee] when we couldn't really understand what it was they [i.e. the applicant] were doing. And I think that would be a great temptation: 'It's something that's not within my field of expertise therefore I'm going to not worry about understanding it. I'm going to rely on someone else to tell me it's okay or not okay.' Because it's *my responsibility* to decide on the ethics of this, not someone else's.

This tension between external reviews and the responsibility of REC members came to the fore when, following the events at Northwick Park, the members of the REC were 'debriefed' by the National Research Ethics Service (NRES), the organisation within the Department of Health that at that point coordinated the REC system:

> subsequently in our debrief after everything happened ... [the point was made that] ... 'it's not our job to assess the safety' ... and the clinicians on the committee were saying, 'well, that actually puts us in quite a difficult position because we have a duty as LREC members but we have a duty of care as clinicians and we have expert knowledge, well, where is the boundary?' ... because we were actually told if we had gone to independent review and an independent reviewer has said it was fine, that would have actually put some liability on our LREC for making a decision saying it was safe, and that that wasn't our remit anyway. So then you end up, you're in a slight grey area as what, what your role is.

The point being made here is that to base a risk–benefit decision upon an external reviewer's assurance of the safety of a drug is to commit the REC to approving the safety of the trial. Yet not only do clinical members of the REC (at least) feel uncomfortable with not considering a drug's safety, whether a drug is considered safe or not is a crucial part of deciding whether the risks of a trial outweigh its potential benefits – along with assessing the information given to research subjects, a core role for RECs. What is being set out by this interviewee with these tensions between different sets of obligations is a clear example of what organisational theorists call 'role conflict' or 'role ambiguity', well documented as a source of impaired decision making.[43]

Perhaps unsurprisingly, the tension that exists at the level of individual REC members and how they deal with external reviews in the context of scientific quality is reflected at the institutional level of the Department of Health. In the case of Phase I trials, one response to the TGN1412 trial has been to set up a specialist expert committee to provide scientific review of complex, high-risk drugs, prior to Phase I trials. Such advice feeds into safety decisions the MHRA makes about clinical trials. The problems with such a solution are underlined in the letter informing RECs of the new advisory group, telling RECs that they:

> may rely on the MHRA to assess the safety of medicinal trials. It [i.e. the REC] is not required to undertake its own safety assessment or seek expert advice on safety issues from expert referees ... although ... For general scientific advice, the REC should either seek further information from the sponsor or consult its own referees.[44]

While such a position – which avoids 'dual' scientific review of Phase I trials by both the MHRA and the REC – makes sense within a framework seeking to ensure the smooth regulation of pharmaceutical trials, it is, with the best will in the world, also internally inconsistent and confusing. It is inconsistent to tell RECs not to make 'safety' decisions, when they are legally obliged by the Directive to weigh the potential risks and benefits of a trial – i.e. make safety decisions. The distinction between 'safety assessments' (where RECs cannot seek independent advice) and 'general scientific advice' (where RECs can consult referees) is unclear and arbitrary; there are many aspects of a scientific review of trial which might reflect upon its safety.

Thus there is clearly a paradox at the heart of the role of science in REC decisions. We know that scientific appraisal plays *some* role in the majority of REC decisions – even more so in those applications that get rejected.[45] We know that very few applications include detailed review of the science[46] – and we might speculate that, perhaps, this is partly why RECs feel the need to offer up their own scientific opinion. The paradox comes with the knowledge that RECs have access to a mechanism that could resolve this tension – commissioning external reviews – which are barely used by these committees.[47]

The data presented in this chapter suggests an explanation for this paradox; commissioning (or, more likely, asking applicants to provide) external reviews involves RECs in trust decisions that are based on very little information (about who does the review, their expertise, and, crucially, their relationship to the applicant). Even for bodies which regularly make decisions about the trustworthiness of researchers, this is a step too far. RECs instead fall back on their own expertise (however limited) rather than make trust decisions in circumstances of such low information. In this sense, an interesting comparison can be drawn with John Downer's work on the Federal Aviation Administration's (FAA's) regulation of the US aviation industry, which argues that modern aircraft are so technologically complex that regulators at the FAA lack the expertise to assess their safety. Instead the regulator uses Designated Engineering Representatives (DERs), who are experienced engineers employed by aircraft manufacturers, to attest to the safety of new designs:

> As employees of the manufacturers, DERs are not sufficiently 'credible' to be the arbiters and guarantors of the knowledge they provide, despite being the only people with the technical competency to provide it. Herein, therefore, lies an epistemic space in which the FAA can work and a function they can perform: they can know the 'experts we do not know'.

In essence the FAA serves as a 'virtuous witness' and:

> attest[s] to the virtue of expert secondary witnesses, such as DERs, and warrant (as independent, publicly accountable actors) that these (potentially biased) experts are worthy of trust. The FAA cannot assess the creditworthiness of technological claims directly, but they can assess the creditworthiness of the people who make them.[48]

RECs, however, cannot carry out a similar role and witness the trustworthiness of the reviewers used by pharmaceutical companies and academic applicants, not least because the lack of information provided about such reviewers renders them opaque and morally unknowable to committees.

This chapter sheds light on the simplistic nature of the standard critique of REC practice around scientific review, the position that argues that

> REC involvement can reflect and be respectful of any prior involvement of other independent experts who have responsibility for such matters. The ethics committee need only engage where they have cause to doubt the ability or independence of the expertise employed.[49]

The point is not that RECs should have cause to doubt, but rather that they lack cause to trust the unknown and opaque external reviewers.

Notes

1 Sarah Edwards, 'The role, remit and function of the research ethics committee – 2. Science and society: The scope of ethics review', *Research Ethics*, 6:2 (June 2010), 60.

2 Department of Health, *Report of the Ad Hoc Advisory Group on the Operation of NHS Research Ethics Committees* (London: The Stationery Office, 2005), 17.

3 Hansard HL Vol. 664, Prt 130 Col GC430 and GC432 (15 September 2004); see also Hansard HL Vol. 661, Prt 86 Col 845–868 (19 May 2004).

4 Department of Health, *Report of the Ad Hoc Advisory Group*, 8.

5 On the relationship between scientific review and lay membership, see: UKNA, MH 160/883, Memo from M.R. Edwards to Mr Morris, ref: H.S.1A A517, 25 August 1967. For a broader analysis see: Adam Hedgecoe, ' "A form of Practical Machinery": The origins of Research Ethics Committees in the UK: 1967–1972', *Medical History*, 53 (2009), 331–350.

6 RCP Archive, Minutes of Royal College of Physicians College Meeting, Thursday 26 July 1984, 23. The initial RCP report on the need for RECs was published in 1967 [Royal College of Physicians of London, *Report of the Committee on the Supervision of the Ethics of Clinical Investigations in Institutions* (London: RCP, 1967)] and was followed up with a second statement six years later [Royal College of Physicians of London, *Report of the Committee on the Supervision of the Ethics of Clinical Investigations in Institutions* (London: RCP, 1973)].

7   RCP Archive, Minutes of Royal College of Physicians College Meeting, Thursday 26 July 1984, 23 and 26.

8   Royal College of Physicians, *Guidelines on the Practice of Ethics Committees in Medical Research* (London: Royal College of Physicians of London, 1984), 2.

9   Royal College of Physicians, *Guidelines on the Practice of Ethics Committees in Medical Research Involving Human Subjects* (London: Royal College of Physicians of London, 1990, second edition), 3–4; Royal College of Physicians, *Guidelines on the Practice of Ethics Committees in Medical Research Involving Human Subjects* (London: Royal College of Physicians of London, 1996, third edition), 8.

10 Department of Health, *Local Research Ethics Committees HSG(91)5* (London: HMSO, 1991), 11.

11 Department of Health, *Governance Arrangements for NHS Research Ethics Committees* (London: Department of Health, 2001), 24. The 2011 revised version of this document makes much the same claim: 'A REC need not reconsider the quality of the science, as this is the responsibility of the sponsor and will have been subject to review by one or more experts in the field (known as "peer review")', Department of Health, *Governance Arrangements for Research Ethics Committees: A Harmonised Edition* (London: The Stationery Office, 2011), 28.

12 George Masterton and Prem Shah, 'How to approach a research ethics committee', *Advances in Psychiatric Treatment*, 12 (2007), 224–225.

13 Richard Barke, 'Balancing uncertain risks and benefits in human subjects research', *Science, Technology, and Human Values*, 34:3 (2009), 356.

14 Sarah Dyer, 'Applying Bioethics: Local Research Ethics Committees and their Ethical Regulation of Medical Research' (PhD thesis, University of London, 2005), 118.

15 Emma Angell, Alan Bryman, Richard Ashcroft, and Mary Dixon-Woods, 'An analysis of decision letters by Research Ethics Committees: The ethics/scientific quality boundary examined', *Quality and Safety in Healthcare*, 17 (2008), 133.

16 'An excessively large study would be unethical if it involved patients continuing to receive a treatment that was already clinically inferior': Andy Vail, 'Experiences of a biostatistician on a UK Research Ethics Committee', *Statistics in Medicine*, 17 (1998), 2813.

17 Department of Health, *Governance Arrangements for NHS Research Ethics Committees* (London: Department of Health, 2001). 15. On several occasions during my fieldwork, the RECs I observed, usually Coastal MREC, brought in a deputy statistician. Historically, prior

to the development of the centralised oversight of RECs within the Department of Health (in around 1999), LRECs were less likely to have a statistician member with one survey listing only 15% of LRECs as having a statistician, as opposed to 100% of MRECs: P. Williamson, J. Hutton, J. Bliss, J. Blunt, M. Campbell, and R. Nicholson, 'Statistical review by research ethics committees', *Journal of the Royal Statistical Society*, 163:Part1 (2000), 5–13.

18 The Medicines for Human Use (Clinical Trials) Regulations 2004 No. 1031, Paragraph 15.5(a), 17.

19 Emma Angell, Alan Bryman, Richard Ashcroft, et al. 'An analysis of decision letters by Research Ethics Committees', 135.

20 Ibid.

21 Richard Barke, 'Balancing uncertain risks and benefits', 357–358.

22 Department of Health, *Governance Arrangements*, 24.

23 Ibid.

24 Department of Health, *Report of the Ad Hoc Advisory Group*, 8.

25 RCP Archive, ms 5062, Minutes of meeting of Committee on Ethical Issues in Medicine, Royal College of Physicians, 13 November 1986, 5–6.

26 RCP Archive, Minutes of a meeting of Chairmen of Ethics Committee held at the Royal College of Physicians of London on Friday 20 February 1987, 4.

27 Richard Barke, 'Balancing uncertain risks and benefits', 350.

28 Minutes of meeting of the National Research Ethics panel, 11 August 2010, 4, https://s3.eu-west-2.amazonaws.com/www.hra.nhs.uk/media/documents/2010-08-11_NREAP_Minutes_FINAL.pdf, last accessed December 2020.

29 David Shaw, 'The ethics committee as ghost author', *Journal of Medical Ethics*, 37:12 (2011), 706.

30 General Medical Council Fitness to Practise Panel (misconduct) case of: Wakefield, Dr Andrew Jeremy, Walker-Smith, Professor John Angus, and Murch, Professor Simon Harry, Day Eight, Wednesday 25 July 2007, 32 and 79.

31 Minutes of meeting of the National Research Ethics panel, 11 August 2010, 4.

32 Emma Angell, Alan Bryman, Richard Ashcroft, et al. 'An analysis of decision letters by Research Ethics Committees', 134.

33 Sarah Edwards, 'The role, remit and function of the Research Ethics Committee', 60.

34 Aaron Kesselheim and Michelle Mello, 'Confidentiality, laws and secrecy in medical research: Improving public access to data on drug safety', *Health Affairs*, 26:2 (2007), 483–491.

35 John Abraham, Julie Sheppard, and Tim Reed, 'Re-thinking transparency and accountability in medicines regulation in the United Kingdom', *British Medical Journal*, 318 (1998), 46–47. For a discussion of the impact of such secrecy on RECs, see: Richard Ashcroft and Naomi Pfeffer, 'Ethics behind closed doors: Do research ethics committees need secrecy?', *British Medical Journal*, 322 (2001), 1294.

36 In informing the teaching-hospital of its increased requirements, the St Swithin's LREC is clearly underlining the highly localised nature of its applicant base; since over 90% of applications came via researchers at St Swithin's Teaching Hospital, this institution was the focus of the Chair's concerns.

37 Mona Shah, Extract from Minutes of meeting of Brent MEC, 23 January 2006.

38 Marcel Kenter and Adam Cohen, 'Establishing risk of human experimentation with drugs: Lessons from TGN1412', *Lancet*, 368 (2006), 1387–1391; Marc Pallardy and Thomas Hünig, 'Primate testing of TGN1412: Right target, wrong cell', *British Journal of Pharmacology*, 161 (2010), 509–511; David Eastwood, Lucy Findlay, Steve Poole, Chris Bird, Meenu Wadhwa, Michael Moore, Chris Burns, Robin Thorpe, and Richard Stebbings, 'Monoclonal antibody TGN1412 trial failure explained by species differences in CD28 expression on CD4+ effector memory T-cells', *British Journal of Pharmacology*, 161 (2010), 512–526; Thomas Hünig, 'The storm has cleared: Lessons from the CD28 superagonist TGN1412 trial', *Nature Reviews Immunology*, 12:5 (2012), 317–318.

39 MHRA (2006) *Clinical Trial Assessment Report Clinical Data* (London: Medicines and Healthcare products Regulatory Agency).

40 Mona Shah, Letter from Brent MEC to Parexel, 31 January 2006; Mona Shah, Extract from Minutes of meeting of Brent MEC, 23 January 2006.

41 Anon, Letter from TeGenero immunologist reviewing TGN1412 preclinical safety data, 9 February 2006; Anon, Letter from immunologist reviewing TGN1412 preclinical safety data, 9 February 2006; Anon, Letter from immunologist reviewing TGN1412 preclinical safety data, 9 February 2006.

42 That time pressures played some role in the Northwick Park disaster is perhaps unsurprising given the evidence that increased speed in drug regulation is associated with higher rates of morbidity and mortality in patients: Mary Olson, 'The risk we bear: The effects of review speed and industry user fees on new drug safety', *Journal of Health Economics*, 27 (2008), 175–200; Cassie Frank, David Himmelstein, Steffie Woolhandler, David Bor, Sidney Wolf, Orlaith Heymann, Leah Zallman, and Karen Lasser, 'Era of faster FDA drug approval has also

seen increased black-box warnings and market withdrawals', *Health Affairs*, 33:8 (2014), 1453–1459.

On the broader international drivers towards faster pharmaceutical review, see: John Abraham and Graham Lewis, 'Harmonising and competing for medicines regulation: How healthy are the European Union's systems of drug approval?', *Social Science and Medicine*, 48 (1999), 1655–1667; John Abraham and Tim Reed, 'Trading risks for markets: The international harmonisation of pharmaceutical regulation', *Health, Risk and Society*, 3:1 (2001), 113–128.

43 See, for example; John Rizzo, Robert House, and Sidney Lirtzman, 'Role conflict and ambiguity in complex organizations', *Administrative Science Quarterly*, 15:2 (1970), 150–163; Robert Miles and William Perreault, 'Organizational role conflict: Its antecedents and consequences', *Organizational Behavior and Human Performance*, 17:1 (1976), 19–44; Gauri Rai, 'Minimizing role conflict and role ambiguity: A virtuous organization approach', *Human Service Organizations, Management, Leadership & Governance*, 40:5 (2016), 508–523.

44 Janet Wisely, 'Letter: First time in man (FTIM) and other clinical trials subject to assessment by the Expert Advisory Group and Commission on Human Medicine', 14 August 2007.

45 Emma Angell, Alan Bryman, Richard Ashcroft, and Mary Dixon-Woods, 'An analysis of decision letters by Research Ethics Committees: The ethics/scientific quality boundary examined', *Quality and Safety in Healthcare*, 17 (2008), 131–136.

46 Minutes of meeting of the National Research Ethics panel, 11 August 2010, 4, https://s3.eu-west-2.amazonaws.com/www.hra.nhs.uk/media/documents/2010-08-11_NREAP_Minutes_FINAL.pdf, last accessed December 2020.

47 Ibid.

48 John Downer, 'Trust and technology: The social foundations of aviation regulation', *The British Journal of Sociology*, 61:1 (2010), 94–95.

49 Stephen Humphreys, Hilary Thomas, and Robyn Martin, 'Science review in Research Ethics Committees: Double jeopardy?' *Research Ethics*, 10:4 (2014), 235.

Conclusion: Regulatory giraffes?

'nothing in the regulatory domain resembles the institutional review board (IRB). To invert the classic story about God delegating authority to a committee to perfect His creations and getting a giraffe in return, the IRB is the giraffe, so odd is it when compared to other creatures in the jungle.'

Harold Edgar and David Rothman[1]

In this book I have set out to explore what kind of things Research Ethics Committees (RECs) are, suggesting that as a kind of regulation they serve to assess and attest to the trustworthiness of researchers, and to legitimate their work. Thus while RECs are interested in the nature of the proposed research – possible harms caused by intervention, issues around capacity to consent, how patients will be recruited, for example – the key insight of this book is that they pay at least as much attention (if not more so) to the (perceived) characteristics of the researchers (and associated staff and organisations), with information gathered through interaction and personal knowledge of applicants, and decision making centring on questions of whether the committee believes that applicants can be trusted to carry out research in the way they say they will, and to protect the interests of participants.

Given that the fieldwork for this book finished well over a decade ago and that, certainly in terms of structures and organisations, the regulation of biomedical research in the UK is characterised by a high rate of change, we might wonder whether the features of the RECs I observed remain in place or my work has become, as Charles Bosk suggests all ethnographies do, 'social history'.[2] In terms of broader structures, in keeping with regular changes to the overall REC system already discussed in this book, there have been steady developments since the end of my fieldwork. At the most banal level, since I left the field the coordinating organisation I dealt with – COREC, the Central Office for Research

Ethics Committees – has been replaced by the National Research Ethics Service (NRES), which was over-seen first by the National Patient Safety Agency (NPSA), and then merged with the newly set-up Health Research Authority (HRA). But, beyond these changes, there has been a steady process of attempts to standardise and centralise REC decision making. We might look at the Shared Ethical Debate (ShED) scheme, piloted in 2007 and fully launched a year later, where a genuine application is sent to a number of RECs for review, with the decision and minutes then 'fed back to the participating RECs, HRA operational teams, the National Research Ethics and Advisors' Panel (NREAP), and the HRA training department in order to develop HRA policies and guidance'. Along similar lines, 2007 saw the launch of an accreditation programme 'to audit RECs against agreed standards', which currently grades committees with 'full accreditation, accreditation with conditions …, or provisional accreditation'. This audit is now accompanied by 'quality control' involving twice-yearly checks on REC practice against agreed standards, and an annual observation of a REC meeting. In 2010, the concept of 'Proportionate Review' was introduced, allowing quicker, sub-committee review of applications judged by researchers to present 'no material ethical issues'.[3]

By the time the Academy for Medical Sciences (AMS) launched its 2010 review of the regulation and governance of health research, there had thus been considerable attempts to standardise REC practice. The initial call for evidence by the AMS sets out a rather gloomy position: 'There is widespread and increasing concern that the process of medical research is being jeopardised by a regulatory and governance framework that has become unnecessarily complex and burdensome.'[4] While this call does not mention RECs by name, the focus of the review on 'clinical trials … experimental medicine, and epidemiological studies' might lead one to expect that revisions to the REC system would be a key focus. Yet the AMS final report largely ignores ethics review, suggesting that: 'NRES and its predecessor, the Central Office for Research Ethics Committees (COREC), have made substantial improvements to the process of ethics review … streamlining regulatory and governance processes in the UK'.[5] The organisation that is currently responsible for co-ordinating RECs in England (and which de facto sets the standards for the other parts of the UK) the HRA, grew out of the AMS recommendation that a single regulator be set up 'to oversee the regulation and governance of health research'[6] and has an expenditure of over £13 million, and employs over 200 staff.[7]

However, I would suggest that, as the historical data explored in this book underlines, while the broader structures and formal rules around REC review might change, the kinds of things that RECs think are important for decision making have remained rather consistent. RECs themselves are substantially the same as the groups of people I observed: the same mixture of experts and lay members, reviewing application forms with numerous questions interpretable in terms of trust, with some (like St Swithin's) drawing heavily on experts from single institutions, a strong role for meeting researchers face to face, and with the same limited engagement with external scientific reviews.

For example, with regard to the REC application form, while question A68 has changed, there are other questions in the current form that provide spaces for RECs to judge applicants' commitment to the ethics review process and hence their trustworthiness: for example A6–2 ('Please summarise the main ethical, legal, or management issues arising from your study and say how you have addressed them'), and A22 ('What are the potential risks and burdens for research participants and how will you minimise them?'). And in terms of localism, despite the attempts to erode local links, there are still RECs with strong institutional ties; based on the most recent annual reports, the 'London – Chelsea REC' draws seven of its nine expert members from the Royal Marsden and Brompton hospitals, or the collocated Institute for Cancer Research. Similarly, the 'London – Hampstead REC' draws six of its 12 experts from the Royal Free Hospital.[8] The attendance of applicants at REC meetings remains central to REC practice, with Edward Dove, based on his recent observations, describing face-to-face meetings between RECs and applicants as a 'vital component' of REC trust in researchers.[9] In addition, data from a freedom of information request shows that, between April 2018 and March 2019, researchers attended REC meetings in 748 out of 880 (85%) applications, in a sample of 20 English RECs (from a total of 65). Finally, with regard to the role of scientific review, it remains the case that 'REC members find it a "constant struggle to try and separate out the idea" that RECs already should be assured that the science is "good" and that the application has had appropriate peer review'.[10] While one might admire the confident assertion of the most recent version of the *GAfREC* document, dated 26 March 2020, that a 'REC will be satisfied with credible assurances that the research has an identified sponsor and that it takes account of appropriate scientific peer review', I cannot help but think that this statement is made more in hope than expectation.[11]

Thinking more broadly, the stories told in this book – about research ethics review as a personal, interactive, resolutely social process – can help us draw conclusions about RECs beyond the specifics of how they reach their decisions. The first key insight provided by this interactional understanding of REC decision making centres on various debates around the need (or not) for RECs to have some kind of ethical 'expert' as a member. The need for 'ethical expertise' on RECs or IRBs is a common claim on the part of commentators, with a typical position arguing that 'If individuals with ethics expertise are not on the IRB ... the IRB would be deprived of their participation and thoughtful contributions to IRB deliberations.'[12] Perhaps the most consistent supporter of such a view is Julian Savulescu, who, in the past, has recommended that 'the requirement to have a religious representative on the [research ethics] committee be replaced with the requirement to have an ethicist', a change necessary on the grounds that 'institutional ethics committees, lacking real moral expertise and knowledge, will slide into subjectivism and ethical relativism'.[13] As justification, Savulescu refers to an 'objective view' of ethics review, where

> there are substantive principles and criteria for the evaluation of research, and the role of the ethics committee is to examine whether the research is ethical by evaluating the degree to which the research conforms to these objective criteria for ethics review.[14]

While it may well be the case that, *in an ideal world*, RECs would make decisions by 'evaluating the degree to which the research conforms to objective criteria', as should be clear from the preceding four empirical chapters, that is not how ethics review currently takes place in the National Health Service (NHS). Given the lived reality of REC review, therefore, and the social, interactional nature of REC decision making, it is not at all clear that there is a need for such committees to have an ethics 'expert' as a member (even assuming such expertise could be identified). Of course, an academic bioethicist may well bring a number of useful skills to a committee – around reading, digesting, and comprehending large amounts of complex information. for example. But these skills are not related to bioethics (a historian of ancient Rome would probably bring the same skill set), but rather their background in academia. Similarly, a bioethicist with a clinical background may well be a very useful REC member, but that would be more to do with clinical expertise than their knowledge of, and ability to apply, abstract philosophical theories. Thus we should be sceptical about claims that ethics committees require people with some kind of expertise in bioethics (or its equivalent) in order to make legitimate decisions.

The second conclusion centres on a topic that has been central to much of the empirical material presented in this book, but dealt with implicitly rather than explicitly, namely how RECs assess the risk of proposed research. Traditionally, research ethics review thinks about risk in terms of the *nature of the intervention* being proposed, and the *characteristics of the population* the intervention is applied to. For example, RECs might view biomedical research that involves the injection of substances or the sampling of body tissue as being more risky than social scientific research that asks people questions. Similarly, people with specific diseases or children or adults who are unable to offer fully informed consent might be viewed as vulnerable populations and hence at higher risk, and in need of greater protection, than other groups. Considerable scholarship has gone into exploring the challenges facing RECs and IRBs in attempting to assess the risks of biomedical research, including: the heterogeneous ways in which the same intervention may impact on different members of the same population;[15] the challenges of evaluating risk in specific research designs (such as comparative-effectiveness research);[16] and the need to distinguish different aspects of decision making around risk – risk–benefit analysis, risk–benefit evaluation, risk treatment, and decision making – from each other.[17]

In the case of UK RECs, while, as we might expect, regulation of risk is part of much broader changes within the regulatory state as a whole,[18] the original *Governance Arrangements for NHS Research Ethics Committees (GAfREC)* document, in place at the time of my fieldwork, was not overly focused on the regulation of risk, simply requiring that, 'Before giving a favourable opinion, the REC should be adequately reassured about … the justification of predictable risks and inconveniences weighed against the anticipated benefits for the research participants, other present and future patients, and the concerned communities.'[19] As Edward Dove maps out, in the ensuing years there has been an increased focus on how RECs think about risk, mainly in terms of encouraging a proportionate response to research felt to offer little risk of harm to patients (such as epidemiological or survey-based research).[20]

However, neither broader scholarly discussion of research risk – including those that offer detailed, step-by-step frameworks for risk–benefit analysis, covering instructions ranging from 'Ensure and Enhance the Study's Social Value' to 'Evaluate and Reduce the Risks to Participants'[21] – nor the formal advice offered to RECs seem to regard researchers, the people who are, after all, going to carry out the intervention being reviewed, as sources of risk in need of assessment. Even those accounts of the role of risk assessment in ethics review that dispute the value of rational choice models and acknowledge the importance of individual character are not talking about the character of the *researcher*, but rather of the potential participant and their personal attitude towards risk.[22]

Yet, as should be clear at this point, failure to take REC decision making about the researcher (or other applicants) into account misses a large part of what these bodies do. There are cases where commentators acknowledge the putative value of such trust decisions in risk assessment, yet they also lament the fact that they are not part of REC custom and practice. For example, Sara Shaw and Geraldine Barrett note in their 2006 review of the UK research governance system and its regulation of risk:

> more complex judgements regarding the character, professional integrity and experiential judgement of the researcher are not explicitly included, though a face-to-face interview at an ethics committee is an opportunity for the researcher to demonstrate these qualities. Arguably, there is a greater need for the formal consideration of researchers' virtues (as well as technical procedures) within risk assessment and governance arrangements generally. Consideration of issues of trust might facilitate risk assessment by allowing committees to explicitly differentiate between different studies and settings.[23]

The key insight of this book is, of course, that while not formally mandated by the guidelines, standard operating procedures, and checklists set out to manage REC decision making, 'complex judgements regarding the character, professional integrity and experiential judgement of the researcher' *are made all the time by RECs.* Indeed, such decisions are central to how RECs go about their business. Seen through the lens of such trust decisions the common complaint that 'IRBs [or RECs] are spending too much time editing informed-consent forms and too little time analysing the risks and potential benefits posed by research'[24] has less bite since looking at consent materials, and how the applicant has completed them, is one aspect of a broader, less asocial sense of risk assessment.

Of course, it may well be that taking trust decisions into account complicates analysis of how ethics review bodies should regulate risk, making it harder to set out rules for how RECs should go about this. Yet any attempt to describe or prescribe the role of ethics review bodies, *without* engaging with the kind of trust decisions they make about researchers (and the factors that underpin those decisions), will inevitably be partial and incomplete.

Finally, the interactional, undeniably social nature of REC decision making allows us to reflect on the nature of ethics review as a form of regulation more broadly. As discussed in the introduction to this book, thinking about RECs as making trust decisions does not fit that well with broader historical changes to UK regulation which, according to Moran, have moved away from the cosy, professional self-regulation ('club regulation') to more formal, centralised approaches. While at the level of organisational rules, with its guidance documents, mandatory training and standard operating procedures, the REC system resembles this modern approach, in terms of actual decisions – centring on interactional assessments of researchers' trustworthiness – RECs are decidedly old-fashioned.

This ambiguous relationship between structural changes to regulatory systems and the nature of the decisions that regulators actually make can be seen elsewhere in medicine, in the case of the regulation of clinical practice in the UK. The standard account – briefly outlined in Chapter 1 – is that, after 150 years of collegial self-regulation, 'the medical profession can now no longer properly be understood as self-regulating'. A series of scandals means that, with regard to the General Medical Council (GMC), the UK's medical licensing organisation:

> The long tradition of doctors both occupying a majority of positions on the GMC as well as controlling the membership of the Council ended in 2009: members are now appointed independently and the Council has parity of lay and registrant (doctor) members.

In terms of the rules for the disciplinary decisions that the GMC makes, the 'standard of proof in "fitness to practise" cases has switched from the high bar used to establish criminal guilt (beyond reasonable doubt) to the lower one used to determine civil liability (balance of probabilities)'.[25] Yet, as Marty Chamberlain cautions, although:

> Some academic commentators, much like doctors themselves, have proclaimed that the reforms introduced ... [have] ... effectively brought an end to the idea that medicine as an autonomous independent profession ... the viewpoint that there has been a decline in medical autonomy is a long standing one which has existed in various forms ever since the emergence of health care managerialism in the NHS in the late 1970s.[26]

His point is that, despite the significant changes made to clinical regulation by the Health and Social Care Act of 2008, 'the issue of the specialist nature of professional expertise, alongside the concurrent need for professionals to exercise discretion in their work, does create a "buffer zone" that protects doctors from outsider surveillance and control'.[27]

For example, despite a considerable increase in complaints against doctors on the medical register (up by 640% between 1995 and 2014), we might be cautious about claims of the loss of self-regulation given that 75% of these complaints are not taken forward by the GMC and 'the shift to a civil [i.e. lower] standard of proof during this time period (i.e. from 2008 onwards) does not appear to have resulted in an immediate and significant increase in doctors being erased from the medical register'.[28] While the reasons for this are complex and cannot be reduced to simple professional self-interest, in exploring cases where the GMC has had to discipline doctors formally convicted of crimes, Chamberlain notes that, in his dataset covering 1,359 cases of convicted doctors referred to

Trust in the system

the GMC, between 2005 and 2015, 'No doctor was barred from practising medicine for serious violent and sex offences, including rape, possession of images of child sexual abuse, manslaughter and domestic violence.'[29]

Again, the causes for such decisions are complex – Chamberlain discusses broader debates around the impact of criminal records on ex-offender rehabilitation, for example – but it is hard to argue with his emphasis on the 'high degree of discretion possessed by the medical profession when deciding if a group member should continue in their employment when they have been convicted of a criminal office'. Given the 'lack of necessary checks and balances needed to ensure public interest and safety needs' the conclusion that 'the medical profession in the United Kingdom appears to be guilty of closing its ranks to protect the privileges of its members at the expense of its public responsibilities' is not an unreasonable one.[30]

In the light of this deep resistance to changes to medical self-regulation *in terms of the actual decisions made*, we can turn our attention back to the regulation of biomedical research by RECs. While, as noted in the introduction, it is important to understand that the researchers regulated by RECs are no longer exclusively, or even largely, members of the medical profession, perhaps we should not be surprised that, despite the trappings of a modern regulatory system, in terms of actual practice, research ethics review looks a lot like old-fashioned club regulation, with a reliance on personal interaction and socially based relationships. Moran's apparent assumption that top-level standards, guidance, oversight, and rules will directly shape the culture and outcomes of regulatory decision making does not seem to apply in the case of REC review.

This discrepancy in the study of regulatory decisions – between the formal level of rules, guidelines, and standard operating procedures and the everyday level at which decisions actually get made – suggests that we need to be cautious in thinking about using broad, system-level analysis of regulatory change as the basis for predicting how individual regulators will make decisions about specific cases. Essentially, it is not at all clear that the conclusions drawn about macro-level studies of regulation (about the demise of club regulation in the UK, for example) will hold when one adopts a micro-sociological approach to regulatory decisions.

In this sense, RECs resemble 'street level bureaucracies', 'hierarchical organizations in which substantial discretion lies with the line agents at the base of the hierarchy'.[31] In his classic work setting out the concept, Michael Lipsky listed street-level bureaucrats as the 'teachers, police officers … social workers, judges', the 'public service workers who interact directly with citizens in the course of their job, and who have substantial discretion in the execution of their work'. Because of these 'high degrees of discretion and relative autonomy from organizational authority', such individuals play an important role in the implementation and creation of policy, including regulatory policy.[32]

While traditional accounts argue that street-level bureaucracies' discretionary powers are a result of limited resources and too high a workload, forcing individuals into making priority decisions for their work, this is not obviously the case for RECs. Here discretion is built into the system through a range of features – members being volunteers, for example, or the relative lack of oversight of the content of REC decisions (as opposed to just the outcome), as well as the historically rooted assumption of the independence of ethics committees – that ensure considerable decision-making autonomy on the part of any individual REC. The key point about the concept of street-level bureaucrats in this context is that their decision making is 'decoupled' from the aims and approaches of top-tier policy makers, and that in the case of research ethics review, this decoupling, and the discretion that goes along with it, open up an important space for trust decision to take place.[33]

I would like to suggest that the stability of the kinds of things that RECs think are important – not just since my fieldwork was finished but also in the preceding decades – is rooted in longstanding 'deep' features of how RECs are organised, both in terms of the task they are required to do (anticipatory review) and the format through which their decisions are made (a multi-disciplinary, hierarchically flat committee). The forward-looking nature of anticipatory review forces REC members to make decisions about researchers' future behaviour, which requires *knowledge* about what they are going to do. Without a crystal ball, the only knowledge REC members have in this context is about the researchers themselves. Making decisions about *what is going to happen*, how specific individuals are *going to behave in the future* and treat other people, requires decisions about trustworthiness, and hence the interactional nature of REC decision making as a crucial source of information to decide on the trustworthiness of the applicant. The form gives insight into the character of the applicant, revealed in more detail through the use of local knowledge and applicants' attendance before a committee. The crucial stumbling block for external scientific review is the unknowability and lack of interaction with reviewers. In terms of the format of REC review, the intensely interactional nature of the committee meeting, with all members carrying the right to speak, underlines the social nature of the format, enhanced by asking applicants to attend meetings to respond to queries.

This perspective takes us beyond a position that argues that the trust-based approach adopted by RECs is a throwback to old-fashioned forms of regulation, rooted in professional self-protection. Instead, it leads us towards the idea that trust-based regulation is thoroughly modern. As Steven Shapin points out when discussing venture capital investment in scientist-entrepreneurs:

> People matter: their personal constitutions matter; their virtues matter. And the reasons they matter has to do with the radical uncertainty of these future-making practices. You need to know about the virtues of people because there is little else you can rely on that is so durable and so salient. While there is a clear link with the premodern modes of familiarity that some social historians and social theorists assure us is 'lost', the reliance on familiarity and the personal virtues is no mere 'survival' of premodernity. Such things don't belong just, or even naturally, to the premodern 'world we have lost'; they belong equally, or even especially, to the world of making the worlds to come.[34]

Of course, trust-based regulatory decisions are not perfect – what human activity is? The decision to approve the TGN1412 trial run at Northwick Park in London, in 2006, is a clear example where trust criteria – a good 'performance' from applicants in front of the committee or assumptions about organisational competence, for example – which normally serve RECS well, failed to identify the risks and problems inherent in the research. This does not necessarily mean that trust is an unsuitable way of regulating biomedical research; there are other approaches on offer, of course, but if one wants to regulate research *before it takes place*, then one has little option but to make trust decisions. To return to the parable offered up by Harold Edgar and David Rothman at the beginning of this chapter, it may well be the case that RECs (or IRBs in their story) are 'regulatory giraffes', but sometimes, you need a long neck.

Notes

1 'The Institutional Review Board and beyond: Future challenges to the ethics of human experimentation,' *The Milbank Quarterly*, 73 (1995), 489–506.

2 Charles Bosk, *All God's Mistakes: Genetic Counseling in a Pediatric Hospital* (Chicago, IL: University of Chicago Press, 1992), 146.

3 Edward Dove, *Regulatory Stewardship of Health Research* (Cheltenham: Edward Elgar, 2019), 56–57.

4 Academy of Medical Sciences, Call for evidence: Review of the regulation and governance of medical research, 2010 https://acmedsci. ac.uk/file-download/34529-129468115986.pdf, last accessed September 2020.

5 Academy of Medical Sciences, *A New Pathway for the Regulation and Governance of Health Research* (London: The Academy of Medical Sciences, 2011), 78.

6 Academy of Medical Sciences, *A New Pathway*, 100. See also: Jean McHale, 'Reforming the regulation of health research in England and Wales: New challenges: New pitfalls', *Journal of Medical Law and Ethics*, 1:1 (2013), 23–42.

7 *Health Research Authority Annual Report and Accounts* 2017/18 (London: HMSO, 2018).

8 The historical institutional roots of these RECs can be traced through their changing names: 'London – Chelsea REC' was formed from a merger of the 'Royal Marsden REC' and the 'Brompton, Harefield & NHLI REC' and called South West London REC 1. The 'London – Hampstead REC' was previously the 'Royal Free Hospital and Medical School Research Ethics Committee' and then North West London REC 2.

9 Edward Dove, *Regulatory Stewardship of Health Research*, 133.

10 Ibid., 135.

11 Department of Health, *Governance Arrangements for NHS Research Ethics Committees* (London: Department of Health, 2020), 26.

12 Laura Beskow, Christine Grady, Ana Iltis, John Sadler, and Benjamin Wilfond, 'Points to consider: The Research Ethics Consultation Service

and the IRB', *IRB: Ethics & Human Research* 31:6 (2009), 6. See also: Nathan Emmerich, 'On the ethics committee: The expert member, the lay member and the absentee ethicist', *Research Ethics Review,* 5:1 (2009), 9–13; Margaret Moon, 'The history and role of institutional review boards', *American Medical Association Journal of Ethics,* 11:4 (2009), 311–321.

13 Julian Savulescu, 'The structure of ethics review: Expert ethics committees and the challenge of voluntary research euthanasia', *Journal of Medical Ethics,* 44 (2018), 491 and 492. Savulescu is specifically writing about ethics committees in Australia where there was a requirement to have such a religious representative on committees.

14 As reported in Merle Spriggs, 'Human subjects research: Review of the NH & MRC national statement on ethical conduct in research involving humans', *Monash Bioethics Review,* 18:4, Ethics Committee Supplement (1999), 12.

15 Michelle Meyer, 'Regulating the production of knowledge: Research risk–benefit analysis and the heterogeneity problem', *Administrative Law Review,* 65 (2013), 237.

16 N. Fernandes, G. Vist, and D. Bryant, 'The concept of risk in comparative effectiveness research', *New England Journal of Medicine,* 372:9 (2015), 883.

17 Rosemarie Bernabe, Ghislaine van Thiel, Jan Raaijmakers and Johannes van Delden, 'The risk–benefit task of research ethics committees: An evaluation of current approaches and the need to incorporate decision studies methods', *BMC Medical Ethics,* 13:6 (2012) https://doi.org/10.1186/1472-6939-13-6

18 Julia Black, 'The role of risk in regulatory processes', in Robert Baldwin, Martin Cave, and Martin Lodge (eds.), *The Oxford Handbook of Regulation* (New York: Oxford University Press, 2010), 302–348.

19 Department of Health, *Governance Arrangements for NHS Research Ethics Committees* (London: Department of Health, 2001), 24.

20 Edward Dove, *Regulatory Stewardship of Health Research,* 88–91.

21 Annette Rid and David Wendler, 'A framework for risk–benefit evaluations in biomedical research', *Kennedy Institute of Ethics Journal,* 21:2 (2011), 141–179.

22 Allison Ross and Nafsika Athanassoulis, 'The role of research ethics committees in making decisions about risk', *HEC Forum,* 26 (2014), 203–224.

23 Sara Shaw and Geraldine Barrett, 'Research governance: Regulating risk and reducing harm?', *Journal of the Royal Society of Medicine,* 99:1 (2006), 14–19; 19.

24 Charles Weijer, 'The ethical analysis of risk,' *Journal of Law, Medicine & Ethics,* 28:4 (2000), 344–361, 344.

25 Mary Dixon-Woods, Karen Yeung, and Charles L. Bosk, 'Why is UK medicine no longer a self-regulating profession? The role of scandals involving "bad apple" doctors,' *Social Science & Medicine*, 73:10 (2011), 1453.

26 John Martyn Chamberlain, 'Reforming medical regulation in the United Kingdom: From restratification to governmentality and beyond', *Medical Sociology Online*, 8:1 (2014), 34.

27 John Martyn Chamberlain, 'Regulating the medical profession: From club governance to stakeholder regulation', *Sociology Compass*, 4:12 (2010), 1035–1042.

28 John Martyn Chamberlain, 'Malpractice, criminality, and medical regulation: Reforming the role of the GMC in fitness to practise panels', *Medical Law Review*, 25:1 (2017), 11.

29 John Martyn Chamberlain, 'Doctoring with conviction: Criminal records and the medical profession', *The British Journal of Criminology*, 58:2 (2018), 394. This category of serious violent and sexual offences was the second largest in Chamberlain's dataset (after vehicle-related offences), making up 18%.

30 John Martyn Chamberlain, 'Doctoring with conviction', 409.

31 Michael Piore, 'Beyond markets: Sociology, street-level bureaucracy, and the management of the public sector', *Regulation & Governance*, 5 (2011), 146.

32 Michael Lipsky, *Street-level Bureaucracy: Dilemmas of the Individual in Public Services* (New York, NY: Russell Sage Foundation, 1980), 3.

33 Previous work addressing regulators as street-level bureaucracies includes: David Hedge, Donald Menzel, and George Williams, 'Regulatory attitudes and behavior: The case of surface mining regulation', *The Western Political Quarterly*, 41:2 (1988), 323–340; Peter May and Robert Wood, 'At the regulatory front lines: Inspectors' enforcement styles and regulatory compliance', *Journal of Public Administration Research and Theory*, 13:2 (2003), 117–139; Roberto Pires, 'Beyond the fear of discretion: Flexibility, performance, and accountability in the management of regulatory bureaucracies', *Regulation & Governance*, 5:1 (2011), 43–69; Suzanne Rutz, Dinah Mathew, Paul Robben, and Antoinette de Bont, 'Enhancing responsiveness and consistency: Comparing the collective use of discretion and discretionary room at inspectorates in England and the Netherlands', *Regulation & Governance*, 11:1 (2017), 81–94.

34 Steven Shapin, *The Scientific Life: A Moral History of a Late Modern Vocation* (Chicago, IL: Chicago University Press, 2008), 303.

Index

Note: 'n' after page reference indicates the number of a note on that page

EU authorised representative for GPSR:
Easy Access System Europe, Mustamäe tee 50,
10621 Tallinn, Estonia
gpsr.requests@easproject.com